THE CRAFT

OF

Handmade Paper

THE CRAFT

OF
Handmade Paper

A PRACTICAL GUIDE TO PAPERMAKING TECHNIQUES

John Plowman

A QUINTET BOOK

First published in Canada by élan press, an imprint of

General Publishing Co. Limited

30 Lesmill Road

Toronto, Canada M3B 2T6

Canadian Cataloguing in Publication Data

is available from the National Library of Canada

ISBN 1-55144-182-9

This book was designed and produced by

Quintet Publishing Limited

6 Blundell Street

London N7 9BH

creative director: **richard dewing**

art director: **clare reynolds**

designer: **simon balley**

senior project editor: **sally green**

editor: **maggi mccormick**

photographer: **paul forrester**

jacket calligraphy: **margaret morgan**

Typeset in Great Britain by

Central Southern Typesetters, Eastbourne

Manufactured in Singapore by

Pica Colour Separation Overseas Pte. Ltd.

Printed in Singapore by

Star Standard Industries (Pte) Ltd

All dimensions are given as

standard (imperial) measurements.

Should you prefer to work in metric, convert inches to

centimetres by multiplying by 2.54.

eg 1 inch x 2.54 = 2.5 cm;

6 inches x 2.54 = 15.2 cm

contents

1

papermaking:
unfolding the story

WHAT IS PAPER? whether a sheet of paper is made industrially or by hand, it will always be the same. Its structure will exploit the properties of cellulose, a natural substance found in all plant fibers. Cellulose, together with fibers containing cellulose reclaimed from rags and waste paper, is the source material of papermaking. Paper is composed of a mass of fibers knitted together. This mass is initially dispersed and suspended in water. It is then collected together to produce a layer on the surface of a mold, through which the water drains away leaving a formed sheet, that is then pressed and dried.

The fibers are bonded together by the cellulose, which expands when it comes in contact with water, thus facilitating their interlocking together.

THE ORIGIN OF PAPER An aspect of the development of any culture, at whatever time, is a need to communicate. People have always sought new ways to acquire, process, and store information. Like the technology causing a revolution now in the latter part of the twentieth century, paper was a major player in another technological revolution, nearly 2,000 years ago.

The early cave paintings communicated aspects of everyday living at that time. As civilizations developed, the spoken language began to be transcribed by means of marks and symbols onto various surfaces. Museums around the world display the writing surfaces of different cultures. These include materials such as wood, stone, metal, bones, and leaves. The relics of ancient Egypt show how hieroglyphs (which was the name given to their script), carved into stone provided a permanent record of their culture.

It was the ancient Egyptians who fabricated a writing surface from the papyrus plant. A sheet of papyrus, from which the word paper is derived, was made by splitting the stalks of the plant lengthwise and soaking them in water. These strips were laid side by side, and another layer was added at right- angles on top. The two layers were then pressed together to form a flat, homogenous sheet. However, although papyrus is a plant material used to form a sheet, the writing surface it creates falls short of being classed as a true paper because it is not broken down to individual fibers.

Parchment and vellum are other surfaces that were produced in ancient times specifically for writing on. Parchment is made with the inner layer of a sheepskin which has been split in two, while vellum is made from the entire skin of a calf, goat, or lamb. Both of these materials have the look and feel of paper, and while papyrus is no longer in use, parchment and vellum are still used today for important decorative documents.

THE DEVELOPMENT OF PAPER

The story of papermaking begins in China. Wood and bamboo were the commonly used writing surfaces, while woven silk was used more sparingly. As the art of calligraphy and the use of block printing developed, there came a need for a writing surface that was easy and economical to produce. So it was that in a.d. 105, Ts'ai Lun produced the first sheet of paper using the fibers of hemp, fishing nets, rags, and barks. Making paper became an integral part of Chinese culture and so it remained until it reached Japan 500 years later. Over the next 500 years, the knowledge of paper manufacture spread through the other countries of the East, finally reaching Morocco. From there, it quickly crossed into Spain, where paper production started in 1150. Over the next 200 years, papermaking mills were set up in all European countries, and by 1690 paper was being produced in parts of North America, with Canada beginning production in 1803.

The source material of papermaking had to be beaten to produce a pulp. This operation was initially done by hand, using equipment based on the pestle and mortar, and evolved into a large, heavy, cantilevered hammer operated by man power. Further developments in the western world followed with the introduction of stamping mills built near a supply of flowing water, necessary for paper production and used as an energy source for the stampers. The stamping mechanism consisted of an arrangement of heavy hammers above a trough filled with the source material. Water power moved the hammers up and down, repeatedly pummeling the material and so separating the fibers. As time went on, stamping mills

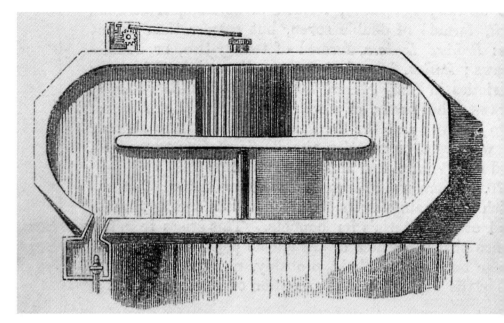

became more efficient and bigger throughout Europe, especially in Germany. However, in Holland, the absence of an abundant running water supply meant that large stamping mill operations were not viable, and an alternative machine, which was more efficient and needed less energy to power it, was invented by the Dutch to separate the fibers from the source material. The source material passed over the base of a box which houses a rotating drum embedded with blades, which chops up the fibers to separate them. The inventor is unknown, but the machine is called the Hollander, and with slight modification, it remains in use today.

The communication revolution ushered in by the invention of the printing press in the fifteenth century had a profound effect on the development of papermaking. Although printing by hand had been well established for many centuries, this mechanization created a new and increased demand for paper, which produced two problems for the papermaker. The first was to supply an expanding industry with sufficient quantities of paper, most of which was still produced by hand. By 1798, this matter was resolved with the invention, by Nicholas-Louis Robert in France, of a papermaking machine based on hand sheetforming methods, but the machine did not become fully operational until the early years of the

19th century, when additional work in England was made by Bryan Donkin, John Gamble, and Henry and Sealy Fourdrinier to the original design. This machine, and a cylinder machine invented by John Dickinson in England in 1809, were based on the principle of forming a layer of matted fibers suspended in water onto the surface of a moving

mold. Paper production was totally revolutionized. An unending length of paper, whose width was determined by that of the machine, could be produced, whereas the dimensions of a sheet of handmade paper were limited by the size of mold that could be used comfortably and efficiently by an individual operative.

The advent of the printing press also increased the urgency to identify a new and abundant source of fiber suitable for papermaking. The dominant source material for making paper was cotton and linen rags, which were sorted and then cut into small pieces and left to rot before being beaten to separate out the fibers. These fibers made a superior paper, but increasing demand made it increasingly difficult to obtain enough cotton and linen rags. Different sources of fiber were located and used to make paper. Eventually the use of paper made from the fibers of wood pulp was strongly advocated, and by the late nineteenth century pulp was in common use.

All these developments have culminated in a paper industry whose product is an essential part of our lives, from the pages of this book to its use in buildings, packaging, paper money, furniture, and clothing. It is amazing that the sophisticated mass production methods used today to manufacture paper are still based on the simple principle invented all those centuries ago in China.

2

basic equipment

The great thing about making a sheet of paper is that most of the equipment is on hand or easily available. To begin, the only specialist pieces of equipment needed are a mold and deckle, and a press. These can be purchased from art and craft suppliers, either separately or as part of a papermaking kit, but since it is straightforward and more economical to make your own, the first part of this book shows you how to make those items that will suffice for the basic papermaking techniques described. If you wish to advance your skills, you may need to purchase additional equipment, but you can often make or adapt items to suit your particular needs. Your working area needs to be near a source of warm water and to be well ventilated. Because paper-

making is an extremely wet activity, wear suitable clothing and protect the work surface and surrounding area with plastic sheeting. All the techniques and projects described on the following pages have been designed to allow you to work in the kitchen, or a garage or home workshop.

The mold, the basic tool of the papermaker, is a frame with a porous surface stretched over it, upon which the fibers are deposited to form the sheet of paper. On top of the mold is placed a deckle, a frame the same size as the mold, that holds the fibers in place. The size of the mold is that of the sheet, the maximum size of which is determined by the ease with which the mold can be handled. In the East, the surface of the mold, known as a laid mold, is made from split

Cutting mat – an essential piece of equipment when cutting dry paper. On the mat, clockwise from left: **scissors** – used to cut dry paper; **utility knife** – used with a ruler for cutting; **pencil** – used to mark out shape to be cut; **poultry baster** – used for pulp painting; **roller** – for rolling wet and dry sheets of paper flat; **pH strips** – essential for testing the acidity of plant material when cooking in an alkaline solution; **ruler** – used when cutting with the utility knife; **Round-nosed pliers** – for bending and shaping wire (when making watermarks etc).

bamboo canes laid next to each other lengthwise across the frame of the mold. The canes are held together by widely spaced threads going across at right-angles, and the parallel lines of the bamboo are then reproduced on the surface of the paper. When the laid mold reached the West, the lengths of bamboo were replaced by lengths of wire held together by thinner wires going across at right-angles.

The other type of mold surface is called a wove mold and consists of a fine wire mesh or gauze stretched across the frame. Initially made from iron, it was quickly supplanted by brass, which is non-corrosive. Although there are slight variations in papermaking in the East and the West, the basic principle remains the same. In Japan the surface of the mold can be separated from the frame, while in the West it is attached to the frame.

There are two ways of forming a sheet. The first involves floating the mold in water and pouring the pulp onto its surface. When the mold is jiggled, the water breaks up the pulp, allowing it to spread to even thickness. The second means dipping the mold into a tub full of pulp and bringing it out so that the paper forms on the mold surface. Once a sheet has been formed, it can either be left on the mold to dry, or the wet sheets can be taken off the mold and stacked. These stacks are then put under pressure to expel the water, and the sheets are separated and left to dry, either by hanging or pasting onto a smooth flat surface.

Disposable kitchen cloth – a felt that is laid between wet formed sheets of paper when couching. **Colander** – this is used to clean plant material as well as pulps. **Liquidizer** – an efficient piece of equipment to beat source material to form a pulp. **Mold and deckle** – the basic tools of papermaking. **Press** – used to expel water from the formed sheets. **Shallow tray** – in which to collect water expelled from the press.

THE ESSENTIAL PIECES OF EQUIPMENT NEEDED TO MAKE A SHEET OF PAPER ARE A MOLD AND DECKLE. THE MOLD IS A FRAME ACROSS WHICH IS STRETCHED A FINE MESH, UPON WHICH A LAYER OF PULP IS FORMED. DIRECTLY ON TOP OF THE MOLD FRAME SITS ANOTHER OF THE SAME DIMENSIONS. CALLED THE DECKLE, IT HOLDS THE

technique

making a mold, deckle, and press

pulp in place on the mold during the sheet-forming process. Although sets are available readymade, making your own will give you a greater empathy with the craft of papermaking. This piece of equipment will serve you well, although frequent use will make the mesh on the mold sag, so it will need to be replaced at intervals. Expanded aluminum mesh is used here, but an alternative would be plain glass curtains; these will need to be replaced more frequently. The size of mold and deckle made here will produce an 8 × 10-inch sheet of paper. Of course the size of the mold and deckle can be adapted, but to start with I would not recommend a frame bigger than this or smaller than 5 × 7 inches. Although a press is not an essential piece of equipment for the papermaker, having a serviceable one certainly makes life a little easier, so we also show how to make a simple but extremely efficient and long-lasting example. They can of course be purchased – a bookbinder's one is ideal – but they are expensive.

"There was an old woman who swallowed a fly": paper pulp piece. **Coo Geller.**

materials

1½-inch screws (non-corrosive)

4 bolts, 4 inches long

8 washers

4 pieces of planed wood,
¾ × 10 inches

4 pieces of planed wood,
¾ × 12 inches

2 pieces of composite board,
¼ × 12 × 16 inches

4 wing nuts

waterproof wood glue

waterproof tape

polyurethane varnish

expanded aluminum mesh

equipment

corner clamp

screwdriver

electric drill

drill bits

countersink bit

paintbrush

staple gun

staples

tape measure

pencil

scissors

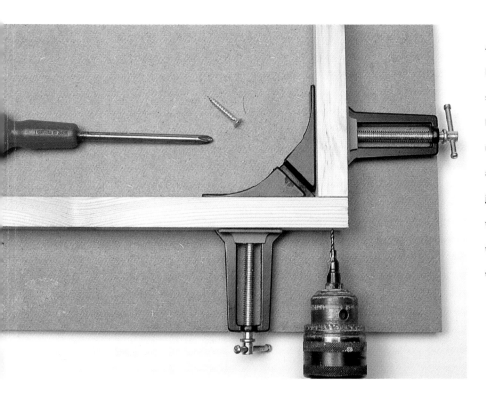

Apply waterproof wood glue to one end of a long length of wood and clamp it at right-angles to a shorter length. After drilling and countersinking a hole, screw the two pieces together to produce a right-angled section. Repeat with the remaining long and short lengths of wood. Use the corner clamp to join these two sections together as before to form the frame. Let the glue dry for 24 hours, then apply two coats of varnish to provide a waterproof finish to the wood.

2

Cut the expanded aluminum mesh so that it is
slightly smaller than the outside dimensions of the
frame. Staple the middle of one side of the mesh,
pull it across the frame, and staple the middle of the
opposite side into the frame; repeat on the other two
sides. Gradually work your way around the frame
stapling the mesh, making sure it is stretched tight
across the frame.

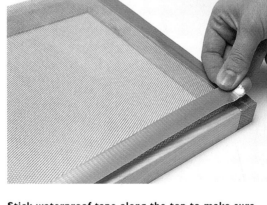

Stick waterproof tape along the top to make sure
that the sharp edges of the aluminum mesh do not
interfere with the sheetforming process.

3

The mold and deckle are now complete. As you can
see, the deckle, made the same way as the frame, fits
neatly on top of the mold.

4

To make the press you need two pieces of board each measuring ¼ × 12 × 16 inches. Drill a hole for a bolt in each corner. Coat both boards with varnish to provide a waterproof finish.

5

The press shown is set up and ready to go. To work efficiently, it must have the holes aligned directly above one another.

6

Collage using dyeing, bleaching, painting. Silk threads and found materials. **Elizabeth Couzins**.

making a pulp

To make a sheet of paper industrially or by hand requires the use of a pulp, a mass of fibers containing cellulose which has been separated from the source material. Fiber and cellulose occur naturally in all plant matter and can be reclaimed from rags and recycled paper. This process of separating fibers is known as beating, done by hitting the source material to break it up and expose the fibers, or mechanically by using a Hollander, a machine designed specifically for this purpose and used both in mass and hand production of paper.

In this section you are shown how to separate the fibers from three types of source materials to produce a pulp. A kitchen liquidizer will beat the source material efficiently as long as you remember to beat only a small amount of material at a time, to add enough warm water (up to the halfway mark should be adequate), and to operate the liquidizer in small bursts. If the motor starts to become labored, there is either too much material, not enough water, or a combination. Take out some material, add more water, and switch on again. It will be quicker to beat recycled paper than plant material, because the plant fibers will tend to wind themselves around the blades of the liquidizer. Each of these sources of fiber requires differing amounts of preparation time. If you use recycled paper first to familiarize yourself with the technique, and then proceed with a part-processed fiber, you will become more confident and should be ready to experiment with plant fibers.

RECYCLING WASTE PAPER TO PROVIDE FIBER FOR PAPERMAKING MAKES ECONOMIC AND ENVIRONMENTAL SENSE. THE QUALITY OF THE PAPER YOU CHOOSE FOR RECYCLING IS INDICATIVE OF THE QUALITY OF THE SHEET OF PAPER TO BE MADE, SO THE LESS PRINT OR INK ON THE PAPER, THE BETTER. AVOID USING GLOSSY

technique

recycled paper

magazine paper; the coating contains chemicals which are not conducive to papermaking. Shredded computer paper, frequently thrown out, is recommended for making paper, as are envelopes, used and new photocopy paper, and wrapping or packaging materials.

materials

shredded paper, preferably non-glossy

equipment

liquidizer

saucepan

colander

1 Shredded paper is ready for recycling. Otherwise, cut or tear the paper into small pieces.

2 Paper with writing or printing should be boiled for 15 minutes to dislodge some of the ink. Then put in a colander and place under running water.

3 Place a small amount of paper into the liquidizer and fill it halfway with water. The action of the liquidizer will quickly separate and disperse the fibers evenly in the water. Repeat the process until all the paper has been reduced to a pulp as shown.

PART-PROCESSED FIBER, WHICH YOU CAN PURCHASE IN SHEET FORM FROM SPECIALIST PAPERMAKER

SUPPLIERS, HAS ALREADY BEEN COOKED AND HAD AN INITIAL BEATING. ALTHOUGH IT IS MORE COSTLY, IT HAS

THE ADDED ADVANTAGE OF BEING TRIED AND TESTED, AND YOU CAN BE SURE THAT A QUALITY SHEET OF PAPER WILL BE MADE

technique

part-processed fibers

from it. The fiber used here, cotton linters, is a versatile fiber ideal for sheetforming

and casting because of its low shrinkage rate.

m a t e r i a l s

cotton linters sheet

e q u i p m e n t

liquidizer

bowl

2

Soak overnight in water to soften and make it easy to liquidize.

Tear a sheet of cotton linters paper into strips of ½ × I inch. It is very easy to tear by hand.

3

Place in the liquidizer, add some warm water, and beat small amounts at a time as before, until they are reduced to a pulp.

PLANT FIBERS PRODUCE A GREAT VARIETY OF COLORED AND TEXTURED SHEETS OF PAPER. THE FIRST STEP IS TO BREAK DOWN THE PLANT MATERIAL BY COOKING IT IN AN ALKALI SOLUTION (ALKALINE SODA IS LESS HARMFUL THAN CAUSTIC SODA). AS A GUIDE, USE 1 OUNCE OF SODA FOR EVERY PINT OF WATER. THE SOLUTION NEEDS TO

technique

plant fibers

have a pH reading of between 10 and 12.

Add enough plant material to be barely covered by the alkali solution, bring to a boil, then simmer. After 2 to 3 hours of cooking, test the pH number of the plant material. If it has a neutral pH number of 6.5 to 7, the cooking is complete. If not, make up another solution and continue cooking until a neutral pH number is obtained. This process will dispose of the unwanted elements of the plant. Once the plant material has been broken down, it can be beaten to separate its fibers and form a pulp. As a general rule the best plants for papermaking have very fibrous stems or leaves; yucca, pampas grass, and nettles are ideal. The pulp can be lightened with bleach if necessary, but use only a small amount as it will affect the quality of the paper. It is important that you wear protective clothing during the cooking process. Rubber gloves and adequate ventilation are essential, and you should protect the surrounding area and avoid splashes.

materials

plant stems

washing soda

equipment

scissors

saucepan or large container

NOT ALUMINUM

liquidizer

colander

netting

pH strips

1 Cut up the plant material (in this case pampas grass) into equal lengths about 2¼ inches long. Then add washing soda to the water in the container, stirring until it is fully dissolved.

2 Add plant material, making sure it remains covered by the alkali solution. Bring to a boil, then simmer for 2 to 3 hours in a well-ventilated room.

3 After boiling, place the mushy plant material in a net and colander and hold it under running water. When the water is clear as it falls from the bottom of the colander, the plant material is clean.

4 Test the pH value of the plant material; a neutral reading of between 6.5 and 7 is required.

If you want longer fibers, mix long and short fibers in the tub. The sheets of paper you form will be stronger. To make longer fibers, place a pile of broken-down plant material on a rigid surface and beat it with a heavy stick or wooden mallet. Although this traditional way is slower, it allows you more control over the length of fiber obtained. **6**

5 Beat the broken-down plant fibers in the liquidizer. To check whether the fibers are short enough after beating, place them in a clear container of water, and check that they float.

GALLERY ONE *pages 24-25*

1 Plant papers ranging from onion to pampas grass.

2 Wild banana paper, made from leaf and plant fibers.

3 Jacqui Jones, "Leaven", molded paper made from leaves.

4 Jacqui Jones, recycled paper.

5 Jacqui Jones, silk fiber paper.

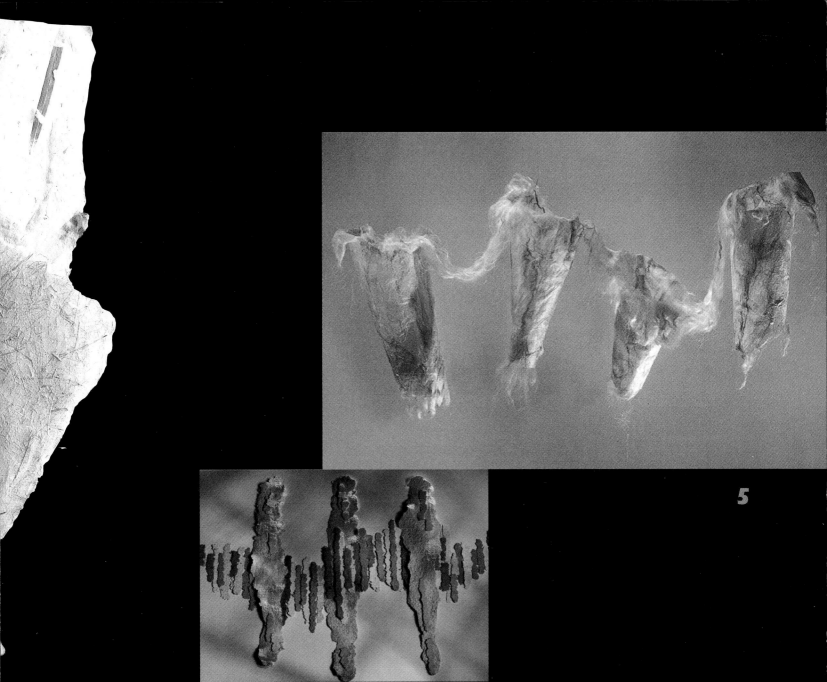

5

making a sheet of paper

Believe me, it is a truly wonderful achievement to have made your first sheet of paper. There is nothing simpler, and it is a technique that is very easy and quick to acquire. Once learned, it is not forgotten. Except for the mold and deckle, the only materials needed are a supply of warm clean water and pulped fibers. Whatever the source material of the pulp, whether you are using plant fiber, part-processed fiber, or recycled paper fiber, the method of forming a sheet remains the same.

When the paper is going to be used for writing, printing, or painting, the surfaces need to be sealed to prevent them from soaking up the ink or paint. This process is known as sizing, and the method shown here is tub sizing, which I feel is most suited to everyday use. The size is added to the tub and mixed with the pulp before the sheet is formed. The

other method, sheet sizing, involves immersing individual sheets of dry paper in size and then allowing the sheet to dry again. As you become more adept at papermaking and start to have a better idea of what you want to do, you can find out more about this aspect of papermaking. The size – a binding agent that can be likened to a glue – used here is gelatin, which can be bought from art material suppliers, but even household gelatin from the supermarket will suffice. Starch can also be used, either that produced by boiling vegetables such as potatoes or the ordinary household variety. Other easily obtainable sizes include white or yellow glue, wallpaper paste, or decorator's size. If you are using recycled paper pulp, especially one where the paper was previously printed or written on, it will already have been sized, so it will not need to be treated again.

THE WESTERN TECHNIQUE FOR SHEETFORMING CONSISTS OF ADDING PULP TO A VAT OF FRESH, CLEAN WATER UNTIL THE FIBER CONTENT IS ABOUT ONE OR TWO PERCENT. THE ACTUAL SHEET IS FORMED BY HOLDING THE MOLD AND DECKLE ABOVE THE VAT, AND DIPPING IT INTO THE SUSPENSION OF FIBERS. THEN WITH A REGULAR,

t e c h n i q u e

sheetforming

continuous movement, the mold and deckle is brought out of the vat, in the horizontal position. A quick shake from side to side allows water to drain through the screen and the pulp to settle into an even surface. The mold is then placed on the side of the vat to drain, before the deckle is removed.

m a t e r i a l s

cotton linters pulp

plant pulp

recycled pulp *(shredded paper)*

gelatin size

e q u i p m e n t

mold and deckle

large plastic tub

measuring cup

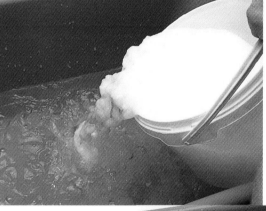

1 As a rule you need enough water in the tub to reach the halfway point on your deckle when it is stood vertically. Add the cotton linters pulp; the amount you add will determine the thickness of the sheet of paper. After a short while you will be able to judge the correct ratio of water to pulp.

2 Following the manufacturer's instructions, mix the gelatin size and then pour it slowly into the tub. Make sure you stir well.

3 It is vitally important that the pulp and water mixture is stirred immediately before you start to form a sheet of paper to give an even distribution of pulp in the water.

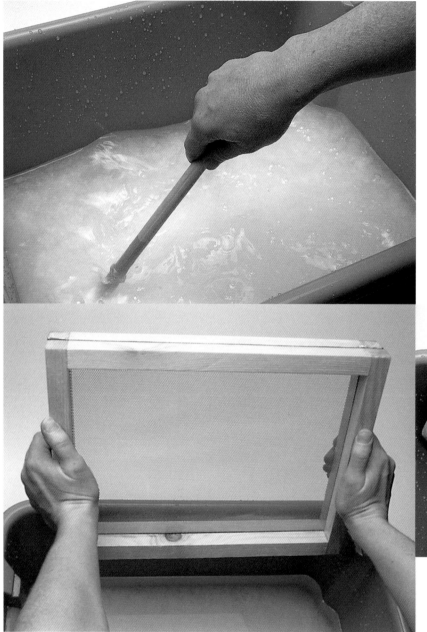

Place the deckle on top of the mold and hold them vertically at arms' length away over the surface of the water. 4

Make sure the pulp and water have been thoroughly stirred, then lower the mold and deckle into the water. In one movement bring the mold and deckle to a horizontal position, 5 making sure they remain beneath the surface.

7 As the speed with which the water drains away slows down, bring the mold and deckle to a diagonal position. Let the last of the water drain into the tub.

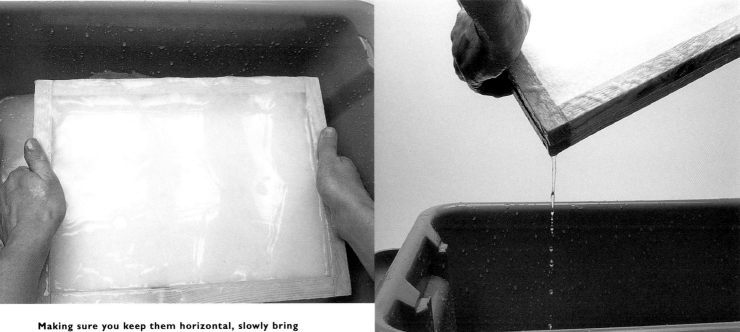

Making sure you keep them horizontal, slowly bring the mold and deckle up through the pulp and break through the surface. Gently shake the mold and deckle back and forth and from side to side. This action gives an even distribution of pulp which in turn means the fibers are crossed over one another to form a tight bond. Keep the mold and deckle in this horizontal position and let the water drain off. **6**

Holding the mold horizontal, carefully lift off the deckle and make sure no water drops onto the surface of the formed sheet on the mold. **8**

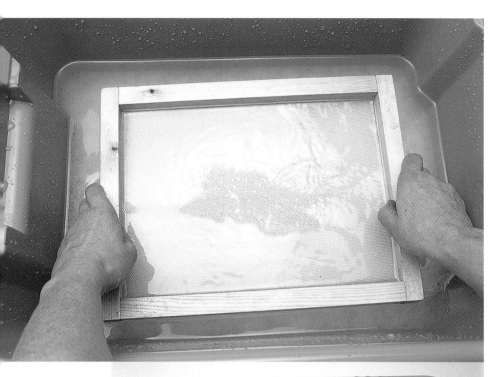

9 If the sheet you have formed is not perfect, turn the mold upside-down and gently bring it into contact with the surface of the water. The formed sheet will come away from the mold. Break the formed sheet up by stirring it back into the pulp and water.

10 Whatever type of pulp you use, the method of forming a sheet is exactly the same. Here a plant pulp has been used. The plant fibers that have collected on the wooden frame as the mold and deckle are brought out of the tub should be carefully wiped away before you remove the deckle.

11 Shredded paper was the source material for this recycled pulp. It is discolored, but this is a desired feature of a piece of paper thus created.

COUCHING IS DERIVED FROM THE FRENCH VERB *COUCHER* MEANING TO LAY DOWN, A PRECISE DESCRIPTION OF THIS TECHNIQUE, WHICH INVOLVES THE TRANSFER OF THE WET FORMED SHEET FROM THE MOLD ONTO A PILE READY FOR PRESSING. IT MIGHT LOOK LIKE A RATHER HAZARDOUS TECHNIQUE, BUT IT IS STRAIGHTFORWARD

technique
couching

and quick to learn. The important point to bear in mind is that it needs to be performed in one continuous movement. Any hesitation means it is more likely that the sheet will not come away from the mold easily. The pile of formed paper is known as a post, and in between each of the wet sheets is a felt, which separates the sheets and soaks up water.

Ostensibly these layers should be made from felt, as the name suggests, but alternatives include interfacing fabric used in dressmaking or disposable kitchen cleaning cloths. Because it is far easier to couch onto a sloping than a flat surface, the first step is to prepare a couching mound, a gently graded slope made of folded wet felts of different sizes laid on the bottom press board. To make sure the formed sheets lie directly and squarely on top of one another, use masking tape to make a registration mark on the bottom press board. This should mark the length of one side of the mold and will aid you when you are positioning the mold prior to couching. It will also be invaluable during the laminating technique (see page 00).

materials

masking tape

disposable kitchen cleaning cloths (*felts*)

heavy old blanket

spring clips

equipment

press

shallow plastic tray

1 Place the bottom of the press in the shallow tray, lay down the blanket and then two felts, each of which has been folded into four. On top lay three felts, each of which has been folded into three.

2 Wet the pile of felts and smooth it down with your hands to form a mound. As an aid to the couching technique, make a registration mark by sticking masking tape along the bottom edge of the board. At each end add two pieces at right-angles to this length. The distance between them must equal the length of the mold.

3 Lay another felt on top and hold it in place with the spring clips. If you smooth the felt down gently with your hands, it will pick up the water from the mound underneath and stick to it.

4 You are now ready to start the couching process, which must be performed in a smooth, continuous movement. If at first you don't succeed, don't worry. Try again. Once you get the hang of it, you won't forget it. Hold the mold containing a formed sheet vertically and position the bottom long edge of the mold inside the registration marks.

5 Holding the bottom of the mold in position, bring it smoothly down onto the couching mound.

At the same time, bring the edge of the mold nearest to you up to finish in a vertical position with the mesh facing you and the formed sheet left on the mound. Cover this sheet with another felt, anchor it with the spring clips, and continue to couch another sheet. Repeat the process to form a post of formed sheets; six is a good number to start with. **7**

6 Once it is on the mound, press down firmly and carefully. You will see that water is then expelled through the mesh of the mold.

UP TO 90 PERCENT OF THE SHEET OF PAPER YOU HAVE MADE WILL BE COMPOSED OF WATER, WHICH NEEDS TO BE EXPELLED AS QUICKLY AS POSSIBLE SO THE PAPER CAN START TO DRY. TO DO THIS, YOU NEED TO USE THE PRESS. PRESSING THE PAPER ENTAILS APPLYING AN EVEN AND CONSISTENT PRESSURE TO THE POST OF PAPER

pressing and drying

which forces the water out and also compacts and strengthens it. It is best to use a press, because you can maintain a consistent pressure and be sure that the post of paper will not move. Another method is to place heavy weights such as bricks or stones on a board placed on top of the post of paper and gradually to add more weights until the last of the water has been expelled.

Alternatively, the post can be placed between two boards and you can hold this "sandwich" between your hands and press to squeeze the water out. This method is not really efficient for a large post composed of large sheets. A better way would be to place the boards on the ground and stand on the top board, so your body weight can apply pressure and force the water out; Make sure you hold onto something. Bear in mind that these last two methods increase the risk of the post of paper moving while it is under pressure. If this happens, it is far more likely that the sheets of paper will be damaged. Once the sheets of paper have been pressed, they can be removed from the post, still on their felts, ready to be dried.

e q u i p m e n t

press

shallow tray

iron

soft brush

roller

When the post is complete, cover with a felt and blanket. Place the other press board on top and turn the whole press over. Remove the top board, lift the felt and remove the couching mound. Replace the heavy felt and board, and turn press back over.

Paper pulp and gesso.
Amanda Goode.

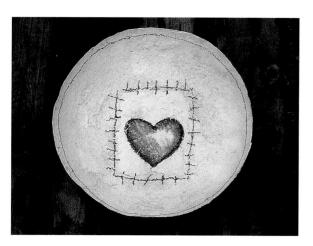

2 With the press positioned over the tray, insert the bolts and start to tighten the wing nuts. Make sure that you do each one evenly, gradually increasing the pressure on the post of paper. Water will then start to be expelled from the side of the press.

As you increase the pressure, the water will really start to gush from the press. Keep turning the nuts to sustain this pressure to expel most of the water. **3**

Remove the bolts, the top of the press, and the top blanket and felt. Be careful not to damage the first sheet. Lift up the felt under this sheet and remove it from the post. While the damp sheet remains on the felt, it is easy to handle without being damaged. Hang the felt on a line so the sheet of paper dries naturally on the felt. This way of drying will produce a wrinkled sheet of paper. **4**

5 To make a flat sheet of paper, lay it carefully, still on the felt, face down on a plastic-coated surface. Brush it down flat to get rid of any air bubbles, and go over the felt with a roller. Once the sheet is flat, carefully peel away the felt. Lift one corner gently and, touching the corner of the sheet, brush it away from the felt and back onto the surface. Peel the felt off slowly and leave the sheet to dry.

7 This sheet of paper has been left on the plastic board to dry. As it is peeled back, you can see the smooth underside. The top is rougher, reflecting the surface of the felt that was used to cover it.

6 To speed up the drying process, lay a sheet down as before and iron on top of the felt to dry the sheet of paper underneath.

8 This sheet of paper shows the wrinkling that is characteristic of being left to dry hanging up.

3

PAPERMAKING VARIATIONS

embedding

The distinguishing feature of an embedded paper is its use of extra material, not part of the pulp, that is held in position by the fibers of the sheet so the embedded material becomes an integral part of the paper. The first method shown is more appropriate for larger pieces of material, which are sandwiched between two sheets using the laminating technique. The registration marks on the press board are used here to make sure the freshly couched sheet lands directly on top of the sheet already couched. This method gives you more control, allowing you to produce more than one sheet containing the same composition. Conversely, when you add material to the pulp as in the second method, there is no control over its positioning, so each sheet of paper will be different, leaving the final design in the lap of the gods! Embedding paper is a great way to provide interesting contrasts, so spend some time deciding what material to embed. If you are using a plant-fiber pulp, why not embed a large part of, or even the complete plant, creating a "before and after" effect. Soak any dry plant material in water first to keep it from floating on top of the pulp. An interesting paper can be created using a recycled-paper pulp embedded with seeds, leaves, or feathers to provide a comparison between the natural and the man-made. An embedded paper is especially suited for decorative uses, such as gift wrap, envelopes, or book covers. However, when the paper is held up to a light source, the embedded material is highlighted, so it can be used with great effect to make a lampshade or a window shade.

THE TECHNIQUE OF LAMINATING (ALSO KNOWN AS "MULTIPLE COUCHING") INVOLVES COUCHING

ONE OR MORE SHEETS OF PAPER ON TOP OF EACH OTHER. THE FIBERS OF EACH LAYER OF PAPER BOND TOGETHER

DURING PRESSING AND DRYING TO CREATE A SINGLE SHEET. YOU CAN USE THIS RELATIVELY SIMPLE TECHNIQUE TO PRODUCE SUBTLY

t e c h n i q u e

embedding
using laminating

interesting effects.

Try using layers of different colors to make a double-sided sheet, or several overlapping layers to create a softer effect. Here, feathers are embedded between two laminated sheets.

m a t e r i a l s

cotton linters pulp

feathers

e q u i p m e n t

mold and deckle

press

I

After couching a sheet of paper, arrange the feathers in a pleasing composition on its surface.

2

Place another sheet of couched paper directly on top. Place a felt, and then the heavy blanket over the sheets and proceed with the pressing operation. Embed and press one sheet at a time; otherwise, you will get impressions of the embedded material on the other sheets as they are pressed.

SMALL PIECES OF SHREDDED FABRIC AND/OR PLANTS CAN BE ADDED TO THE PULP TO CREATE BEAUTIFUL AND UNUSUAL PAPERS. MAKE SURE YOU CHECK THE COLOR FASTNESS OF ANY MATERIAL USED, IF YOU DO NOT WANT COLORS TO MERGE WITHIN YOUR PAPER. ALTHOUGH THE PAPER USED HERE CAN BE COUCHED AND PRESSED

technique

adding to the pulp

in the normal way, if you are using larger elements in your pulp, dry the paper on a board after pressing. This will stop it from wrinkling around the embedded elements.

materials

colored acetate pieces

pulp

equipment

tub

mold and deckle

press

1 Sprinkle the acetate pieces evenly into the pulp.

2 As the sheet of paper is formed and the mold brought out of the tub, the acetate pieces are trapped in with the pulp. Couch and press this sheet in the normal way. With this method of embedding, you can couch and press a post of sheets.

EVERYTHING INCORPORATED INTO THIS ATTRACTIVE HANGING IS NATURAL, PAYING HOMAGE TO THE

ENVIRONMENT, AND COMBINES THE LAMINATING AND EMBEDDING TECHNIQUES THAT HAVE BEEN COVERED SO

FAR. THE STRING IS FUNCTIONAL AS WELL AS DECORATIVE, HELPING TO HOLD THE SHEETS OF PAPER TOGETHER. SHEETS OF

project

fringed and embedded hanging

different fibers are joined together
without glue, providing an excellent
illustration of the strength and efficiency
of the natural cellulose contained in
plant fibers as a bonding agent.

materials

plant papers

string

dried flowers

equipment

**laminated board bigger than the
finished size of the hanging**

roller

palette knife

felts

techniques

embedding

sheetforming

laminating

1 Lay three different sheets of plant paper, pressed but not dried, lengthwise on the laminated board, overlapping their ends. Lay a felt on the overlap and go over it with a roller to press the sheets together. Repeat with the other two overlaps. On one side, position a piece of string lengthwise and form a loop at the top. Then return and lay the string lengthwise to the bottom. Repeat on the other side. Take a selection of dried flowers and arrange them between the strings for a pleasing composition.

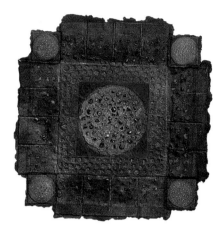

Handmade collage using dyeing, bleaching, painting, with silk threads and found materials. **Elizabeth Couzins.**

3 Fold back the top of the onion paper, carefully push dried flowers in between these two sheets, and press the sheets back together tightly around the flowers. Arrange dried flowers along the top of the hanging to complete the composition.

2 After pressing, lay a sheet of onion paper directly on top of the bottom sheet. When it is in position, carefully lay a felt over the top and roll it firmly to make sure that it sticks to the sheet underneath.

4

To hold the loose material in place, tear off a strip of yucca paper and lay it carefully across the middle of the top sheet to entrap the string. Lay a felt over the top and carefully roll in order to press the strip onto the sheet underneath.

5

Lay another strip of yucca paper on top of the strings, pushing the paper down so it molds itself around the string. Lay a felt on top and go over it with a roller. Continue this procedure at intervals to make sure the string is held firmly in place. Leave the hanging to dry on the laminated board.

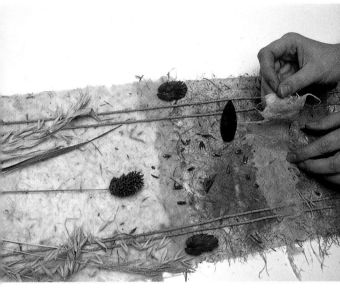

6

When the hanging is dry, slide a palette knife between the hanging and the board. Work your way slowly around, then carefully pull the hanging off the board. Hang using the embedded strings.

1

3

2

4

5

6

TWO

embossing

An embossed paper is one in which pressure has been used to leave the impression of an object on the surface of a sheet of paper. The whole or part of the surface of the sheet of paper can be embossed to create interesting shapes and textures. Because the paper used in this technique is thicker, add extra pulp to the tub before starting to form a sheet. Choose an item to emboss that has a well-defined shape and surface, and avoid objects whose surface is very finely textured or decorated.

Embossed papers are ideal to use as wall hangings or covers of handmade books, where a thicker paper is required. Any loose surface on the embossed object, such as paint, dirt, grease, or rust, will be left on the surface of the paper, so unless this is the desired effect, clean the object thoroughly before embossing. When embossing fresh leaves or other plant material, color will bleed out onto the paper during the pressing process. This can be a virtue of the embossed paper, but if this effect is not wanted, use only pressed and dried plant material. Because the fibers of papers made from plants are generally larger and bulky, it is more difficult to emboss them. If a plant paper is used, make sure it has been made using short fibers.

PAPER CAN BE EMBOSSED EITHER BY FORMING IT ON A TEXTURED SURFACE, OR, AS HERE, BY PRESSING A TEXTURED SURFACE AGAINST IT. BY VIRTUE OF THE PAPER USED IN THIS TECHNIQUE BEING THICKER, THE WIRE MESH CAN BE PRESSED FIRMLY DOWN INTO THE PAPER. THE PRESSURE OF THE RELIEF ELEMENT APPLIED AGAINST IT QUITE

technique

using a wire mesh

dramatically alters the surface of the paper to form a satisfying textured pattern. For more subtle embossing

patterns, you could try pressing a dishcloth, tea towel, piece of lace, or embroidered cloth onto the paper, remembering to cover it first with a felt.

materials

formed sheet of paper

wire mesh

equipment

press

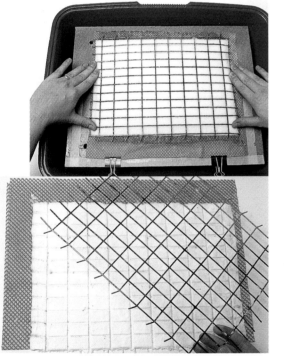

1 Couch a thick sheet of paper and place the wire mesh on top, pushing well into the surface. Lay a felt and a heavy blanket over the top and press. Leave the layers in the press for at least four hours, with the wing nuts tightened.

2 Remove the sheet and mesh and let the paper dry on the felt. The paper will shrink, so remove the mesh before the paper is fully dry, leaving an impression of the mesh on the surface of the sheet.

THIS ATTRACTIVE WALL HANGING SHOWS THE EMBOSSING TECHNIQUE BEING USED TO PROVIDE A JUXTAPOSITION OF THE NATURAL (EMBOSSED LEAVES) AND THE MAN-MADE (RECYCLED PAPER). THE LAMINATION TECHNIQUE IS USED TO JOIN PART OF EACH SHEET TO THE NEXT, SO THIS HANGING CLEARLY DEMONSTRATES THE

Japanese wall hanging

strength and durability of seemingly tenuous connections. The four sheets come together in an unusual configuration, reminiscent of a kimono, and the use of bamboo as a hanging device is in keeping with the oriental look of this hanging.

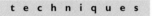

materials

- recycled paper-pulp sheets
- leaves
- string
- bamboo cane

equipment

- laminated board
- roller
- palette knife
- brush

techniques

- sheetforming
- laminating
- embossing

1 Couch a thick piece of paper onto the press and position a leaf on top. Cover with a felt and heavy blanket, and press.

2 After pressing, lay the sheet flat on the laminated board. Overlap the end of another wet sheet on the first one. Lay a felt on the overlap and go over it with a roller, pressing the two sheets together.

3 Emboss two more sheets with leaves as before. Lay them diagonally across the opposite top corners of the top sheet and brush them into position. Lay a felt on top and roll over each corner smoothly, pressing the sheets together firmly.

4 Starting at the bottom, lay a length of string across the sheets, encircling each of the embossed elements. Don't worry too much about controlling the string as it is laid down – the more haphazard, the better. Once the string is in place, lay a felt over the top and firmly and smoothly roll over the top to embed it firmly into the paper.

5 Lay a length of bamboo cane across the top corners of the hanging. Carefully fold these corners over and around the bamboo, tucking it underneath and pressing the paper layers firmly together. Leave the whole hanging flat to dry.

6 Once the hanging is dry, carefully remove each of the leaves to reveal an impression in the surface of the paper. Notice that some of the color from the leaf has stained the paper. Finally take the string off the paper, which leaves a very distinct impression behind.

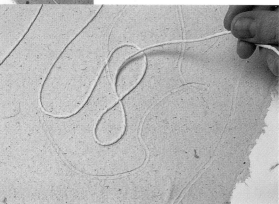

coloring

Natural plant fibers used in papermaking can themselves produce a wonderful array of subtle color. But if you want to experiment more, an exciting variety of color and texture opens up to you. In the early days of papermaking, an assortment of natural substances was used to color paper, ranging from berries to woad.

Later, the color of paper came to reflect its quality — best quality rags only were reserved for white paper, colored rags produced tougher paper more suitable for wrapping etc.

It wasn't until the 18th century that dyes started to be added to the pulp. Penetrating the structure of the fibers, they have a natural affinity with cellulose. Dyes available today include: organic dyes, usually derived from coal tars (called direct dyes); natural dyes, derived from plant sources; and dyes that form a chemical bond with the fibers. The latter are most often used in the textile industry. You can also experiment with using spices for dyeing. Paprika, coriander, and cumin produce earthy, organic shades, and turmeric rich yellows.

Here we are going to look at coloring pulp and marbling, but of course papermakers today are using color in ever more exciting and varied ways. Sheets of different colored wet pulps can be overlapped to produce stunning effects, or different colored layers placed directly on top of one another. Color can also be applied afterwards; adding oil paint onto handmade paper, for example, adds a wealth of contrast.

YOU CAN GET SOME REALLY EXCITING COLORS BY ADDING A DYE TO THE PULP BEFORE YOU START TO FORM A SHEET OF PAPER. COLD-WATER FABRIC DYES ARE EASY TO OBTAIN AND USE, OR YOU CAN PURCHASE SPECIALIST DYES SPECIFICALLY FOR COLORING PAPER. WHATEVER DYE YOU CHOOSE, ALWAYS MAKE SURE THAT YOU

technique

coloring pulp

follow the manufacturer's instructions and recommendations. Make sure that the color is fast and will not run, especially when laminating wet sheets of colored paper. There are many ways to color pulp, and it is worth experimenting. Try using tea or coffee, the juice from berries, food colorings, and spices. In the following examples, cotton linters pulp is used, pure white and perfect for coloring. Always wear rubber gloves and/or barrier cream when using dye.

materials

pulp

cold-water fabric dyes

equipment

rubber gloves

plastic container

fine net

colander

mold and deckle

plastic tub

barrier cream

Following the manufacturer's instructions, mix the dye and pour it into the pulp. Mix well to obtain a consistent color.

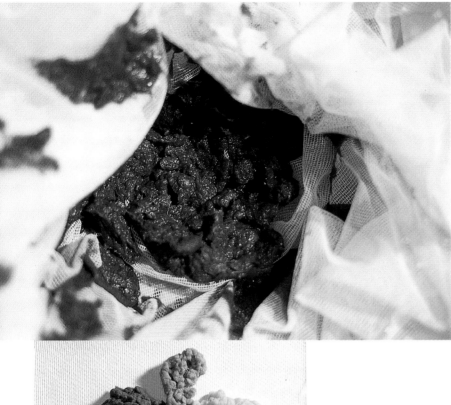

2 Leave the pulp to soak in the dye for a while, then pour it into a colander lined with fine net. Run water over and through the pulp to wash away excess dye.

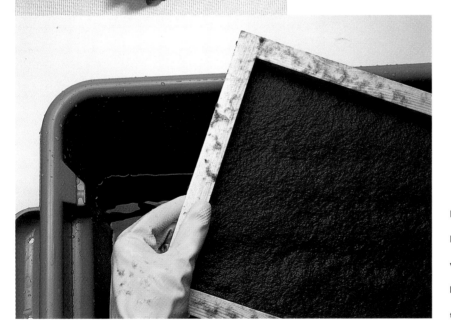

3 Start with the bold primary colors – red, blue, and yellow – and then mix them in order to obtain an exciting range of colors.

4 Place the colored pulp in the tub and mix it well. Form the sheet, and press and dry it in the normal way. Make sure you wear rubber gloves and/or a barrier cream to prevent your hands from becoming the same color as the sheet of paper.

MARBLING IS ANOTHER WAY TO COLOR HANDMADE PAPER, BUT THIS METHOD IS USED AFTER THE

PAPER HAS BEEN MADE. IT IS A RELATIVELY SIMPLE AND STRAIGHTFORWARD DECORATIVE TECHNIQUE

THAT ORIGINATED IN PERSIA IN THE SIXTEENTH CENTURY WHICH ALLOWS YOU TO GET SOME REALLY AMAZING EFFECTS.

technique

marbling

Marbling is based on the fact that oil and water do not mix, so an oil-based color will float on the surface of water. The colors can be mixed into patterns that can be picked up on paper, but cannot be reproduced exactly.

materials

oil paints

wallpaper paste

ox gall

handmade paper

newspaper

mineral spirits

equipment

shallow tray

small dish

small paintbrush

dropper

knitting needle or toothpick

Squeeze some oil paint into a small dish.

1

Add a small quantity of mineral spirits and then mix well together.

2

3 To test for the correct consistency, dip a brush into the mixture and hold it above the dish. The paint should slowly form a drop and fall off. If the paint drops too fast, add more paint. If it doesn't drop at all, add more solvent.

5 Add a drop or two of ox gall to the paint mixture to aid the dispersion of the paint across the surface of the water.

4 To thicken the water, add a small amount of pre-mixed wallpaper paste and stir it in well.

6 To test whether you have added the right amount of ox gall, drop some paint onto the surface of some water in a small container. It should immediately spread across the surface of the water, forming interesting patterns. If it fails to spread, add more ox gall to the paint mixture

Repeat steps 1 to 6 to mix all the colors you wish to use. Drop paint onto the surface of the water in the test bowl. It should spread satisfactorily and give an indication of how well the colors look together.

7

Dragging a strip of folded newspaper across the surface of the water will break its surface tension and should be done before starting the marbling technique. The same method is used to clean the surface of the water after marbling. As you drag the strip across the water toward you, the paper picks up the paint remaining on the surface of the water.

8

Hold the brush loaded with paint horizontally above the surface of the water and tap it gently so that drops of paint fall onto and disperse across the surface of the water. Change position and continue dropping paint over other areas of the surface.

9

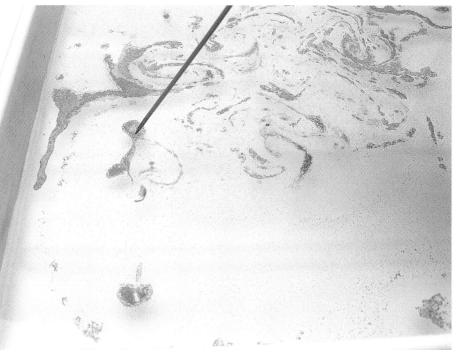

10 Repeat the process to add another paint color, then use the point of a knitting needle or toothpick to move the areas of paint into one another until you have created a pleasing and interesting design.

11 Holding a sheet of paper at diagonal corners, lay it carefully onto the surface of the water. Start with the corner farthest away from you, then slowly allow the paper to lie on the surface of the water. The paper should be removed when the water begins to soak through the back.

Carefully hold the corners of the paper and lift it up vertically from the surface of the water in one smooth, continuous movement. Lay the sheet flat on top of a thick pad of old newspapers or an old towel and let it dry. The resulting dry sheets can then be ironed in order to flatten them. **12**

13

For a more controlled pattern, try the traditional marbling technique known as "combing". Drop two colors on top of the water, and with a knitting needle or toothpick, drag through the water along the length of the tray to produce parallel lines.

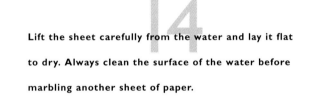

14

Lift the sheet carefully from the water and lay it flat to dry. Always clean the surface of the water before marbling another sheet of paper.

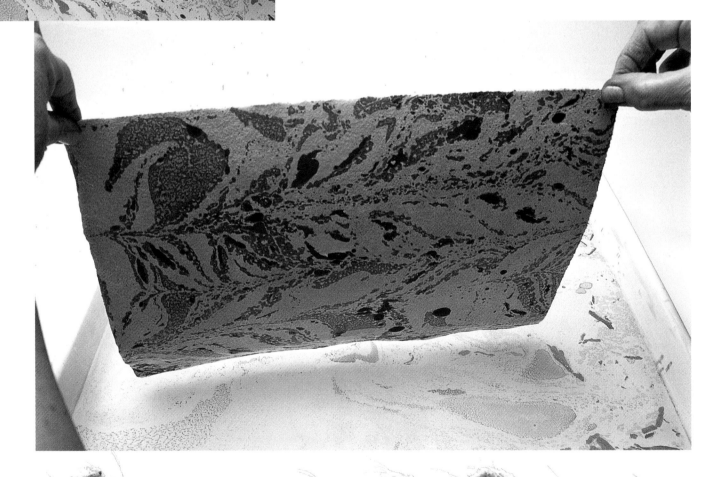

IN THIS PROJECT YOU CAN EXPLORE AND EXPERIMENT WITH MAKING DIFFERENT COLORED PULPS

TO PRODUCE BRIGHTLY COLORED SHEETS OF PAPER THAT CAN BE USED TO MAKE A VIVID WALL HANGING. IN

ADDITION TO USING THE TECHNIQUES OF LAMINATING AND EMBEDDING, YOU ARE SHOWN HOW TO REMOVE THE MALLEABLE PULP

project

brightly
colored hanging

as part of the composition. It is possible to make

effective and stunning patterns simply by scratch-

ing your design into the wet pulp. Once you have

mastered this technique, you will be able to employ

it to ever greater effect in your papermaking.

materials

blue, yellow, and green pulp

2 yards of ¾-inch-wide
    orange ribbon

string

masking tape

equipment

mold and deckle

tub

roller

press

soft brush

rubber gloves/barrier cream

laminated board

techniques

sheetforming

laminating

embedding

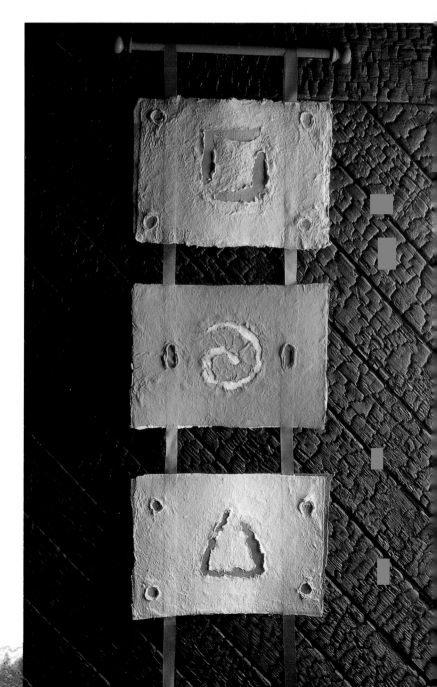

I

Anchor a length of string to the laminated board. Using a brush, lay down two blue and one yellow sheet of paper, going over each with a roller before removing the felts. Make sure they are evenly spaced and align the short edges with the string. Two lengths of ribbon, each I yard long, are placed on top and ¼ inch in from either side of the sheets, so that it traverses all three sheets with an even length overhanging at each end.

2

Using green pulp, form another sheet and remove the deckle. Using a finger, gently push into the sheet to mark a square shape, then delicately remove this area of pulp from the mold. Couch and press as usual. Using the same technique, make a yellow sheet with a triangular shape and a blue sheet with a spiral.

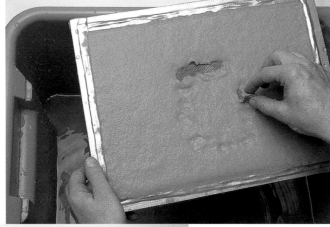

3

Carefully laminate each sheet onto a sheet as shown to embed the two lengths of ribbon. The color of each bottom sheet shows through the revealed area of the top sheet. Once all the top sheets have been laminated, lay a felt over them and use a roller to laminate the top and bottom sheets together.

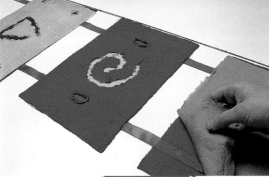

4

As a final decorative and compositional finish, gently scratch away some of the wet pulp of each top sheet of paper to reveal the color of the sheet (or ribbon, as in the middle section) underneath. Add a hanging rod at either end to complete the project.

shaping paper

So far we have dealt only with paper created using a traditionally-shaped mold and deckle. It is possible, however, to create different shaped paper by changing the shape of the deckle, so that the pulp fills up a different shape. To create a simple, circular shape, you could use frames with woven meshes that are ready-to-hand – an embroidery frame would be perfect for the job. Alternatively, you could make a shaped deckle using styrofoam board to create interesting shapes. Cake molds and other kitchen equipment such as pastry cutters also make good shaped deckles. You could use shapes made from deckles such as these to put on the front of handmade cards, just to give one example.

Papermakers make and use shapes in a variety of ways. Several shaped papers can be tied together and then carved, cut, or added to before being painted etc. Other artists create interesting shapes by keeping the sheets unrestrained during couching and then allowing them to fall on to a base, thus reflecting the natural qualities of wet pulp. Pulp can also be poured through stencils to create amazing shapes. These can then be used to create collages. Shaped paper made from different pulps can also make for very interesting textured collages. It can be torn to provide a contrast between the smoother lines created by the shaped deckle and the rough ones created by tearing by hand, which reveals more of the fibers.

Here we cover using shaped deckles to make an envelope, a lampshade, and a collage.

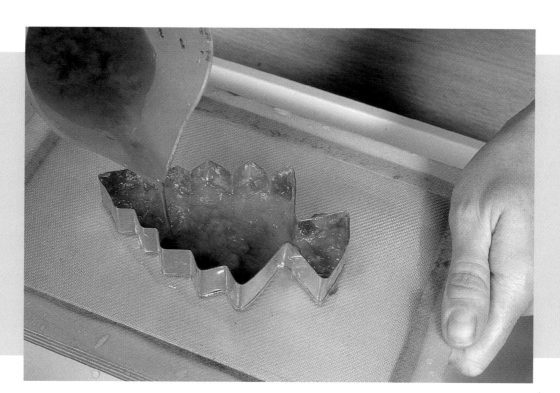

INSTEAD OF CUTTING A DRY PIECE OF PAPER TO THE DESIRED SHAPE, IT IS POSSIBLE TO MAKE SHAPED SHEETS OF PAPER BY USING A SHAPED DECKLE. THIS METHOD OF MAKING A SHAPED DECKLE IS VERY SIMPLE, USING EASY-TO-CUT STYROFOAM BOARD. AS LONG AS THE STYROFOAM IS WELL WATERPROOFED, SUCH A DECKLE WILL

technique

shaped deckle

last a reasonable amount of time, but is not permanent. To make a more durable shaped deckle, you need to use a material such as ¼-inch masonite, plywood, or composite board cut out using an electric jigsaw.

materials

¼-inch styrofoam board the
same size as the frame of
the deckle

envelope

varnish

equipment

cutting mat

pencil

steel ruler

utility knife

paintbrush

Place the deckle on top of the styrofoam board and draw around the inside edge to correspond to the limits of the mold. Open out the envelope, place it within this rectangle, and draw around it.

Use the utility knife to cut out the shape to make a square board with a shaped hole in the middle.

2

Paint both sides and all the edges with two coats of varnish to waterproof the board, letting it dry between coats. Place the deckle on the mold and form a shaped sheet of paper in the usual way.

3

THE SHAPED DECKLE CAN BE USED TO FORM A SHEET WHICH CAN BE MADE INTO AN ENVELOPE ONCE IT IS PRESSED AND DRIED, ALLOWING THE SHAPED PAPER TO BE PUT TO A PRACTICAL RATHER THAN A DECORATIVE USE. THE ENVELOPE COMPLEMENTS THE SHEET OF WATERMARKED PAPER MADE ON PAGE 00, CREATING A STATIONERY

THE ENVELOPE COMPLEMENTS THE SHEET OF WATERMARKED PAPER MADE ON PAGE 00,

project

envelope

set that is ideal to use for special occasions. When you use any shaped deckle to form a sheet, it must be held firmly in place to make sure that the inside edges of the shape are kept tight against the mesh of the mold to keep pulp from going in between the deckle and mesh, which would result in an imperfectly shaped sheet.

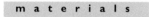

materials

- pulp
- white glue
- double-sided tape

equipment

- tub
- press
- mold and shaped deckle
- cutting mat
- ruler
- brush
- bone folder

1

Hold the shaped deckle firmly in place on top of the mold and proceed with the sheetforming process. When you bring the mold and deckle out of the tub, there will be pulp deposited on top of the deckle. Carefully wipe it off into the tub. After all the water has drained away, remove the deckle. Take care not to drip water onto the formed sheet.

2

Couch the shaped sheet in the normal way. Press the mold firmly down on the couching mound. Bring the mold up vertically to reveal the shaped paper deposited onto the couching mound. Continue making sheets to produce a post of about six sheets. Then press and dry them as usual.

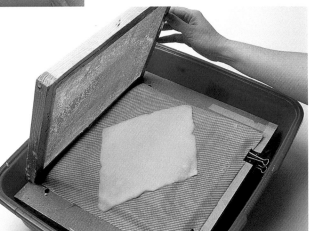

3

To fold and crease each flap, place a ruler along the line of the fold, using the recessed V shapes as a guide. Bring the flap up vertical and fold it over the ruler. Crease the fold with the bone folder by running it along the length of the fold.

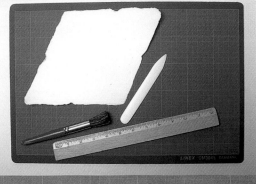

4

Using white glue, brush glue onto the underside edge of three of the flaps. Push them down firmly and glue them in position. Add a length of double-sided tape under the top flap so this flap can be sealed when the envelope is filled and ready to use.

AN EMBEDDED PAPER IS IDEAL TO USE FOR MAKING A LAMPSHADE, BECAUSE THE EMBEDDED

MATERIAL CAN BE SEEN SHINING THROUGH THE PAPER. A READY-MADE LAMPSHADE FRAME IS COVERED WITH

EMBEDDED SHAPED PAPER WHICH IS COATED ON BOTH SIDES WITH A FLAMEPROOFER FORMULATED FOR USE ON PAPER. ALTHOUGH

project

lampshade

any size frame can be used, forethought needs to be given to the shape

of paper to be made and to the ease with which it can be attached to

the frame. The frame used here is a standard shape which divides into

six identical sections. Six sheets of this shape are made, each embedded with short lengths of thread. This lampshade should

not be used with a bulb of more than 60 watts.

materials

lampshade frame

embedded shaped paper

white glue

techniques

embedding

shaped paper

gluing

equipment

styrofoam board

cutting mat

pencil

pulp

cut threads

mold and shaped
  deckle

brush

utility knife

flameproofer spray

1 Position the frame in the center of the board. Draw around the outside of one section, rolling the frame over to complete the outline of the section. Draw a ½-inch wide margin at the top, bottom, and one side of the outline of the section. Cut this shape out of the board with the utility knife.

2 The cutout shape will be used as a template. Cut off the three ½-inch margins with the knife.

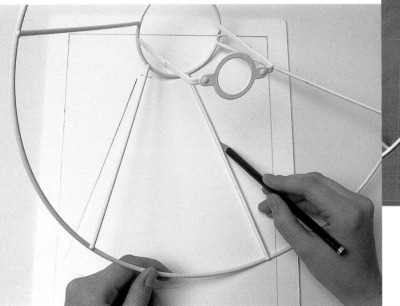

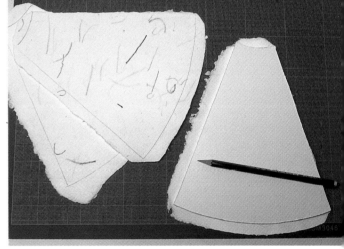

3 Using the shaped deckle, form six sheets of shaped paper embedded with short lengths of thread and let them dry. Place the template on each dry sheet, lining it up with the left-hand side. Making sure there is an equal margin at the top and bottom, draw around the template to reinstate the ½-inch margins on each sheet of paper.

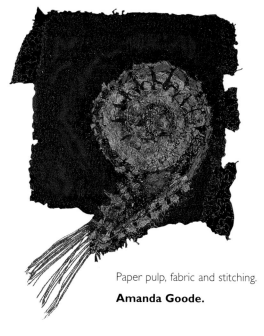

4

Apply glue carefully inside the drawn line of the side margin. Position the side of another sheet on top and butt it up to the drawn line; press together firmly. Repeat with the other sheets to obtain a circular shape with just one edge left to be glued. Let the glue dry.

Paper pulp, fabric and stitching.
Amanda Goode.

5

Place this circular shape around the frame with the pencil marks on the underside. Pull the shape tightly and evenly around the frame and glue the last overlap together. Press together firmly and let the glue dry before proceeding. Apply glue to the inside of the top margins and fold them over. Press them down firmly and let the glue dry. Spray both sides of the paper with a flameproofer suitable for paper.

COLLAGE IS A COMPOSITION USUALLY MADE OF PAPER SHAPES, SOMETIMES COMBINED WITH OTHER MATERIALS, WHICH ARE GLUED TO A BACKING SURFACE. USUALLY THOUGHT OF AS A DRY CRAFT, THIS PROJECT SHOWS HOW NEWLY FORMED AND PRESSED WET SHEETS OF PAPER CAN BE USED TO MAKE A COLLAGE. THE NATURAL CELLULOSE

collaged city landscape

in the fibers anchors the shapes onto the wet backing sheet. This project offers a good opportunity to experience the flexibility and resilience of a wet sheet of paper. Use bold, uncomplicated shapes when composing the collage. A good example is the uncomplicated profile of a head in a silhouette portrait. The profiles of buildings also provide an interesting composition. The cityscape in this project has been derived from photographs of city skylines, using the buildings in the photographs as the source of the shapes for the collage.

materials

styrofoam board

colored pulps

recycled-paper pulps

techniques

collage

laminating

shaped deckle

sheetforming

equipment

mold and shaped deckles

ruler

utility knife

brush

varnish

press

roller

Place a template on a piece of board the same size as
the mold frame, and draw around it. Use the utility
knife to cut out the shape to make a shaped deckle.
Repeat with the other three templates.

3

Two base sheets of recycled pulp paper have been laid down with the short side of one overlapping the long side of the other and rolled together. On top of the base sheet, lay down each of the colored shapes. They will still be wet after pressing, but can be lifted and moved into position if handled with care.

2

To waterproof each shaped deckle, apply two coats of acrylic varnish to both sides and all the edges.

4

Place a felt over the colored shape and brush it into position, making sure there are no air bubbles. Go over the felt with a roller, pressing it into the base sheet for a good bond.

pulp painting

It may surprise you to learn that you can actually paint with paper. Pulp painting is an exciting way to produce a picture by adding colored pulps to the surface of the mold, allowing the freedom to experiment. It uses the same principle as sheet-forming. The cellulose in the fibers bond together to form a sheet, except that the pulp painting is not pressed, but left on the mold to dry resulting in a rougher surface texture. The mold is then out of action until the paper is dry, so if possible have more than one mold. When drying pulp paintings, let as much water as possible drain away; then lean the mold against a wall or other vertical surface and let it dry. When the painting is completely dry, use a palette knife to prise up the edges of the painting gently all the way around and slowly peel the painting up from the surface of the mold. It is helpful to push the underside of the mold gently to help free the pulp painting from the mesh. The colored pulps used here originated from colored paper napkins soaked in water and beaten in the liquidizer. This is an inexpensive source of colored pulps, particularly when small quantities are required. These pulps can be used with confidence since the dyes have already been set so the colors in the pulps will not run together.

Pulp can also be spattered, sprayed, dipped, poured – there is no end to its malleability. For papermakers today, pulp painting can be combined with other techniques to create beautiful works of art. In addition to the methods we cover here, plastic squeeze bottles can be used to contain the pulp for painting. When you have got more confidence, try applying the pulp using spontaneous, freehand strokes.

DIVIDE THE SURFACE OF THE MOLD INTO SIMPLE SHAPES USING STRIPS OF CARDBOARD. EACH AREA IS FILLED WITH A COLORED PULP TO CREATE THIS DECORATIVE DESIGN. THIS METHOD OF PULP PAINTING ALLOWS YOU TO MAKE A PRECISE DESIGN WHICH CAN BE REPEATED SO THAT MORE THAN ONE SHEET OF PAPER OF THE SAME

technique

pulp painting

design can be made. For instance, sheets of the same design can be combined to make a frieze going around the top of walls, whose coloring could reflect the color scheme of the room. Alternatively, a single sheet could be used to decorate and complement an otherwise dreary book cover, as here.

The mold and deckle are placed in a shallow tray with enough water to make sure that the surface of the mold is just below the water. The mold and deckle are held in position by two weights. Submersing the mold in this way suspends the pulp in water after it has been added and means it is evenly distributed when the mold is lifted out of the water.

materials

colored pulps

thin cardboard strips

equipment

mold and deckle

shallow tray

stones or weights

spoon

Place the mold and deckle in a shallow tray half filled with water. Push the mold down and hold it in place with weights placed on two opposing corners. This will also expel any air trapped between the mesh and the water.

2 Place a thin strip of cardboard across the mold a third of the way down the frame. If the strip is slightly longer than the width of the mold, it can be wedged in place. Take a longer length, position each end in the two opposite corners of the horizontal strip to create a curve, and wedge it in place. On top of the horizontal strip arrange three V shapes so that one is centered with another on each side in a corresponding position, creating seven areas.

3 Using the spoon, carefully fill the area within the curved shape with blue pulp. Hold the spoon just above the surface of the mold and let the pulp drop slowly off the spoon onto the mesh, allowing it to settle to an even thickness. Continue by filling the alternate sections above this blue shape with yellow pulp. Gently jiggle the mold so that the pulp is evenly distributed.

4 As the mold fills with pulp, take extra care to avoid dropping pulp into already filled areas. Finally fill the lowest section with the darker red pulp.

5 Carefully lift the mold and deckle out of the water and let most of the water drain off. Position it on the side of the tray, and carefully lift out the strips of cardboard. Make sure you do not dislodge the pulp. After removing the deckle, use a spoon to fill in any areas of exposed mesh with pulp. If necessary use the spoon to spread the pulp out. Leave the painting to dry on the mold, when it is dry, remove it from the mold.

CREATING AN ILLUSION OF SPACE ON A FLAT SURFACE IS THE BASIS OF CONVENTIONAL PAINTING. HERE THE FORMED SHEET OF PAPER ON THE MOLD CAN BE COMPARED TO A BLANK CANVAS. THE COMBINATION OF SIMPLE PERSPECTIVE, THE DIFFERENT COLORS AND TYPES OF PULPS, AND THE IMAGINATIVE USE OF LEAVES WHICH TAKE ON THE

project

landscape

role of trees produce a landscape that has a real sense of depth and atmosphere. This landscape has not been drawn or painted on the surface of the paper, but actually becomes an integral part of it, highlighting how pulp painting can be used to embed other material, in this case leaves, in paper. Before starting, look at landscape painting to become familiar with the conventions and possibilities of this art form. Because you are using a base sheet, take great care when painting with the pulp. Don't drop it from too great a height, as the wet base sheet is easily damaged.

materials

colored pulps

leaves

formed sheet of paper

equipment

mold

shallow tray

spoon

knife

techniques

sheetforming

pulp painting

embedding

1

Form a sheet of paper, remove the deckle, and place the mold over a shallow tray. Have the colored pulps ready to use.

2

Hold the spoon close to the surface of the sheet and drop small amounts of blue pulp onto the formed sheet to create the impression of sky.

3

Sloping hills are made on each side of the sheet, one using green pulp and the other straw pulp. Areas of red pulp are added to provide a variety of tone.

4

Drop some straw pulp onto part of the leaf to embed it in the paper. Let the water drain away, then lean the mold and painting diagonally against a wall to dry.

When the whole painting is dry, go around the edges carefully with a knife and prise up the edges, then gently peel the whole painting away from the mold.

5

THE COOKIE CUTTER AND POULTRY BASTER, AS WELL AS BEING ESSENTIAL PIECES OF KITCHEN EQUIPMENT, CAN BOTH PLAY A PART IN THE MAKING OF A PULP PAINTING. A COOKIE CUTTER, AVAILABLE IN A VARIETY OF SHAPES AND SIZES, CAN BE USED AS A STENCIL, ALLOWING A CONTROLLED APPLICATION OF THE PULP. THE POULTRY

stencil and freehand design

baster is used for the freehand application of pulp, but its advantage is that it allows control over the direction, positioning, and amount of pulp applied. Familiarize yourself with the use of the baster before you start a pulp painting. Draw up pulp into the baster, making sure the pulp is diluted with water. If it is too thick, it will be difficult to draw up and it will clog the baster when it is squeezed out. Do not squeeze too hard; the force of the pulp coming out of the end can blow away the pulp already formed on the mold surface.

materials

colored pulps
formed sheet

equipment

shallow tray
poultry baster
mold
cookie cutter

techniques

sheetforming
pulp painting

I

Form a sheet, remove the deckle, and position the mold over a shallow tray. Select a cookie cutter with a strong profile and prepare the colored pulps in dishes which you should place nearby.

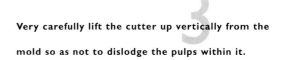

3

Very carefully lift the cutter up vertically from the mold so as not to dislodge the pulps within it.

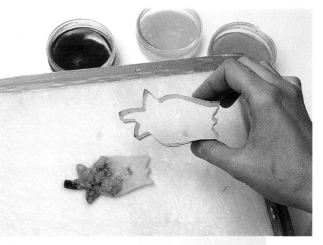

2

Position the cookie cutter carefully in the center of the sheet, taking care not to dislodge the pulp. Fill the baster with green pulp and fill the top of the stencil, then gently squeeze yellow pulp into the bottom of the stencil. Finally add a touch of red.

Squeeze out red pulp to make a frame around the flower shape. Fill the baster with blue pulp and make some evenly spaced short diagonal lines across the red frame. When you use the baster don't squeeze too hard; otherwise, the pulp on the surface of the mold will be blown away.

4

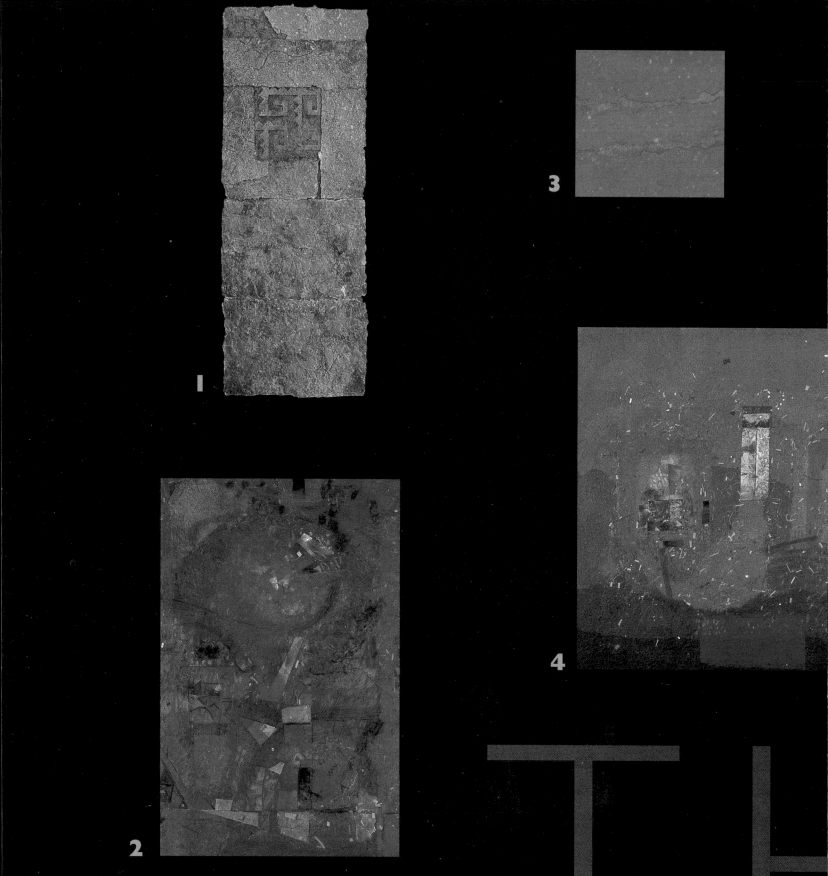

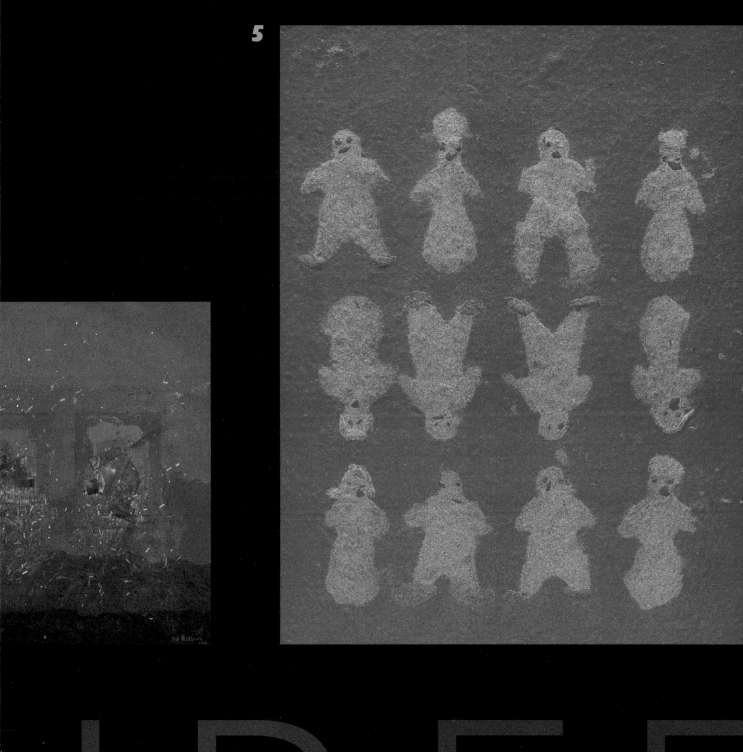

watermarks

Making a watermark is a time-honored tradition dating back to 13th-century Italy. Although it is called a "water"mark, the term is somewhat misleading, since it is not actually water that is responsible for creating the image. Earlier terms such as "wiremark" and "papermark" are more appropriate.

A watermark is in fact a translucent area in a sheet of paper, which usually only becomes visible when held up to the light. This is accomplished by making an impression in the pulp during sheetforming, usually by means of a fine wire shaped into the design you wish to achieve. This wire is attached to the surface of the mold. Instead of covering the surface in an even fashion as usual, the pulp settles over the raised surface of the wire in a slightly thinner layer.

Images such as humans, animals, crosses, stars, and circles were originally used. It is easy to imagine that watermarks in those days might have been invested with a magical or even religious significance. Nowadays they tend to be used for logos and trademarks. Watermarks can of course be made using methods other than the traditional wire. Here we cover making watermarks using self-adhesive plastic tape, which can be cut into any shape you want, and using a thread pulled through two laminated sheets. Once you have the knack of making a watermark, you can start to experiment. Adhesive labels, for example, which can be bought from stationers, come in all shapes and sizes, and can be used to achieve interesting watermark effects.

REGARDLESS OF ITS DESIGN, A WATERMARK IS A THINNER SECTION OF THE SHEET OF PAPER WHICH WILL ONLY BECOME APPARENT WHEN THE SHEET IS HELD UP TO THE LIGHT. THE TECHNIQUE WHICH INVOLVES LAMINATING AN INCOMPLETE SHEET ONTO A FULLY FORMED ONE, SHOWN HERE, IS A VERY GOOD INTRODUCTION TO THIS PROCESS.

technique

making a watermark

To make a thinner sheet of paper, the deckle is not used with the mold. Consequently, the sheet of paper thus produced will have a thinner area, visible when it is held up to the light. The sheet is made using a self-adhesive plastic shape, which provides an area of resistance when it is stuck to the surface of the mold. When the mold is brought out of the tub, the pulp does not adhere to the area covered with plastic, and an exposed area is left on the formed sheet.

m a t e r i a l s

self-adhesive plastic

pulp

e q u i p m e n t

mold

roller

pencil

scissors

tub

press

Lay the roller on top of the self-adhesive plastic and draw around it, then cut this shape out.

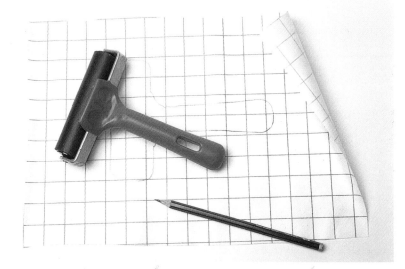

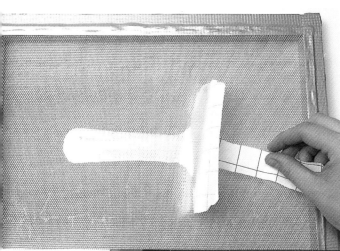

2 Peel away the backing, then position the shape onto the center of the mold and stick it down. Turn the mold over and lay it flat. Roll over the mesh to make sure the plastic is firmly stuck.

3 Form the sheet of paper in the normal way. As it is brought out of the tub, the plastic shape acts as a resistance to the pulp and can be easily seen. Notice that no deckle is being used.

4 The incomplete sheet is laminated on top of a fully formed sheet. Use the registration marks to make sure it falls directly on top of the couched sheet.

5 The incomplete sheet clearly shows the cut-out shape in the middle. Press and dry the two sheets together to complete the lamination technique to produce a sheet with a thinner area.

THIS PROJECT IS A VARIATION ON THE THEME OF MAKING A SHEET OF PAPER CONTAINING THINNER

AREAS, MAKING VISIBLE WHAT WAS ONCE INVISIBLE. IT USES A COMBINATION OF TECHNIQUES, ALREADY COVERED,

IN A VERY IMAGINATIVE WAY. AREAS OF COLOR AND LENGTHS OF THREAD ARE EMBEDDED USING A LAMINATING TECHNIQUE, AND

technique

using a thread

the thread is then used to make the top sheet thinner, revealing what has been embedded between these sheets. The colors that are embedded between the sheets become more apparent when the sheet is held up to the light.

materials

formed sheets

colored pulps

thread

equipment

spoon

mold and deckle

felts

1 Carefully lay down a formed sheet that has been couched and pressed, keeping it flat. Add a variety of colored pulps to the surface of this sheet. Remember not to drop them too far.

2 Continue adding colored pulps until the entire surface of the sheet is completely covered in a haphazard way. Lay three lengths of thread lengthwise over the top of the pulp, letting them overhang at each end of the sheet.

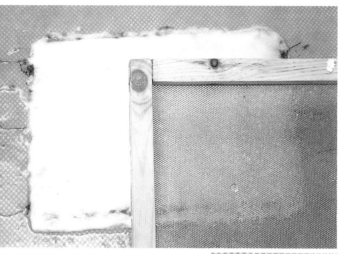

3 Couch another sheet of paper on top. This can be a bit tricky because you will not have the benefit of a couching mound. Try to line the two sheets up as best you can, but it is not absolutely essential that they are laid directly on top of one another.

4 Now gently and carefully take the end of one of the lengths of thread and slowly pull it back. As it tears the top paper, the colored pulps underneath are revealed. Continue to pull the thread back along the length of the paper, and repeat with the two other lengths of thread. Leave the sheet to dry flat on the felt; this will take longer than usual because of the thickness of the paper.

5 Once the sheet is dry, the curled-up edges of the torn strips provide a tactile frame for the colors that are revealed.

WATERMARKS CAN BE VERY SOPHISTICATED AFFAIRS. YOU ONLY HAVE TO HOLD A PIECE OF PAPER MONEY UP TO THE LIGHT TO SEE THE COMPLICATED IMAGERY THAT CAN BE OBTAINED. THIS PROJECT SHOWS HOW TO MAKE A SIMPLE LINEAR WATERMARK. IT COULD BE INITIALS OR A MONOGRAM CONTAINING THE DESIGN OF INITIALS. THIS

project

making a letterhead

personalized paper can be used as stationery in conjunction with the envelope made on page 00.

In keeping with tradition, this linear watermark is made with a length of wire that is permanently attached to the mesh of the mold with thread. Whenever this mold is used, it will form sheets containing a watermark. Because the wire is so thin, the pulp lies on top of it during the sheetforming technique and produces a thinner area of pulp which replicates the design of the wire. Thin electrical wire is an alternative to the jeweler's wire used here.

materials

silver wire

thread

pulp

equipment

mold

pliers

needle

tub

press

scissors

small file

techniques

sheetforming

making a watermark

couching

pressing

drying

1

Using a continuous line, draw a design on paper. Hold the wire over the design and use the pliers to bend it into a shape that follows the drawn line. Use a small file to blunt the two ends of the wire to prevent these jagged edges from damaging the formed sheet.

3

Remember when you begin to form a sheet not to use the deckle, because a thinner sheet is needed. As the mold is brought out of the tub, the wire design is covered with pulp. The uneven edges of the sheet occur when the deckle is not used.

Place the wire shape centrally and toward the top of the mesh on the mold. With the needle and thread, come up through the mesh close to one side of the wire, take the thread over the wire, and push it back down through the mesh on the other side of the wire. Tightly tie both ends of the thread together underneath. Repeat at evenly spaced points along the length of the wire shape.

2

When couching the sheet, make sure it is pressed well down on the couching mound. As the mold is brought vertical, the impression of the wire shape can be seen on the sheet on the couching mound, while the wire design is left on the mold. Form a post of watermarked paper, and press and dry it in the normal way.

4

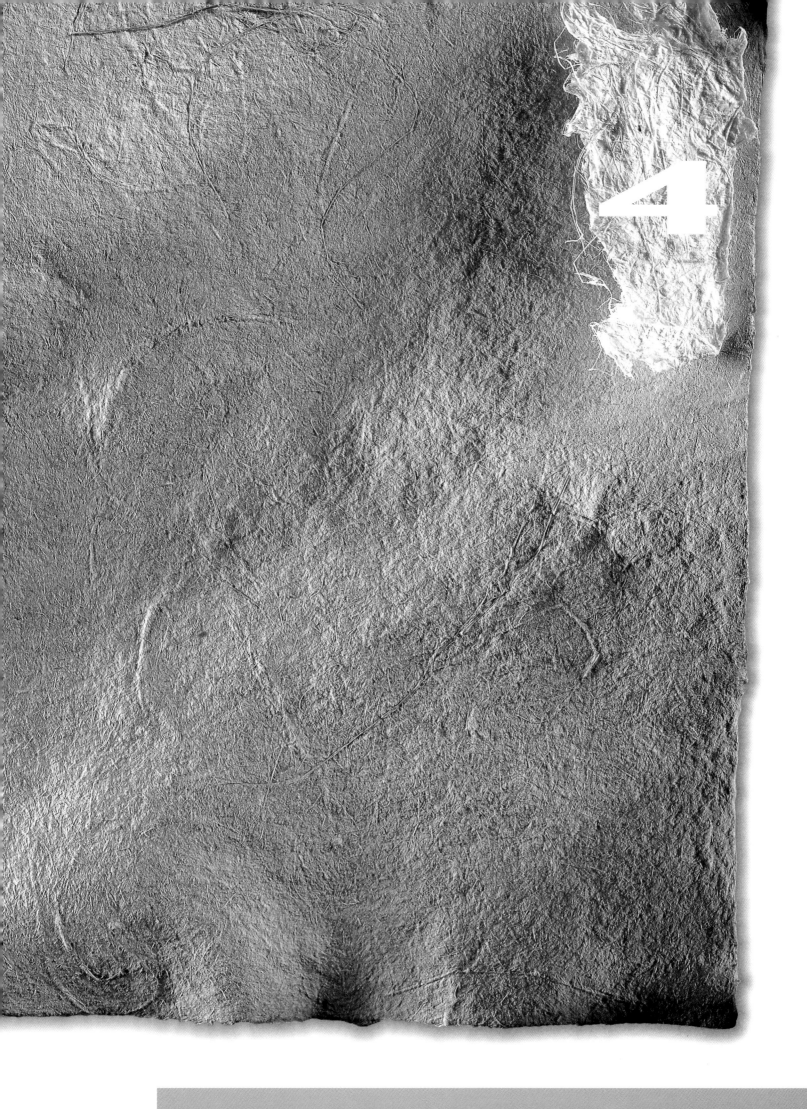

SCULPTING WITH PAPER

casting

While everybody is familiar with papier-mâché, the art of using small pieces of paper with glue and size, etc. in order to create three-dimensional objects, the huge potential of paper's sculptural qualities has really only come to light with the advent of casting. Paper is an extremely flexible medium, and can be used either in its pulp form, in strips of pressed but not dried paper, or even whole sheets, to fill a vast variety of casts and achieve "sculptured" objects. Paper's beauty in this technique lies in the fact that it can achieve almost any shape, yet remain light.

Bear in mind, though, that your choice of pulp can be of paramount importance depending on what you are aiming to make. Different fibers have different characteristics, but the main rule to remember is that the longer the beating (particularly of fibers such as flax) the higher the shrinkage quality of the resulting pulp. There is also a greater degree of translucency to that achieved by a shorter fiber such as cotton linters. When you become more accomplished, you can play and experiment with these different levels of shrinkage, but the techniques described here all employ cotton linters pulp for minimum shrinkage. Accomplished papermakers can control and manipulate shrinkage during the drying period, and sometimes allow it to occur freely, resulting in some quite dramatic forms.

When you start, it's also best to use the most simple casts available and ready-to-hand – there are many manufactured as well as natural objects ideal for the purpose. Papermakers today are using an inspired range of objects, from spades to sand.

CASTING IS THE TECHNIQUE OF PRODUCING A COPY OF AN ORIGINAL. THIS COPY, KNOWN AS THE CAST, IS USUALLY MADE OF A DIFFERENT MATERIAL FROM THE ORIGINAL. THE CAST CAN BE MADE (TAKEN) DIRECTLY FROM AN ORIGINAL OR FROM A NEGATIVE IMPRESSION THAT IS MADE OF THE ORIGINAL OBJECT, THE MOLD. TAKING A

technique

two basic casting techniques

look at gelatin molds illustrates this concept perfectly; the gelatin is the positive image of the negative shape in the interior of the mold. Blister packaging, a familiar sight in stores and supermarkets, provides a vast selection of interesting shapes that can be used as molds to provide an introduction to the casting technique. There are two methods of using paper as the casting material. One uses pulp, while in the other, strips are torn from wet formed sheets. Both methods rely on cellulose to bond the fibers and hold the cast together as it dries. While a cast can theoretically be taken from anything, once it is dry, it must be easy to lift the cast away unhindered, and it is vital that consideration is given to this fact before a cast is made. A releasing agent, in this case petroleum jelly, is applied to the surface of the original to prevent the paper from sticking. Another point to bear in mind is that it is desirable to take the cast from an interior rather than exterior surface of an object, because paper shrinks as it dries. The shape of a cast on an interior surface will not be constricted as it shrinks, whereas the shape of a cast on an exterior surface will crack and break as it shrinks. It is therefore advisable when casting to use a pulp that has a minimum of shrinkage, such as cotton linters.

materials

blister packaging

petroleum jelly

colored pulps

formed sheet

equipment

rubber gloves

brush

sponge

1

When selecting the blister pack, choose one from
which the cast can be easily removed.

3

Wearing gloves, pick up a small quantity of pulp and
squeeze some of the water from it. Don't squeeze
too hard or the pulp will be too dry to work.

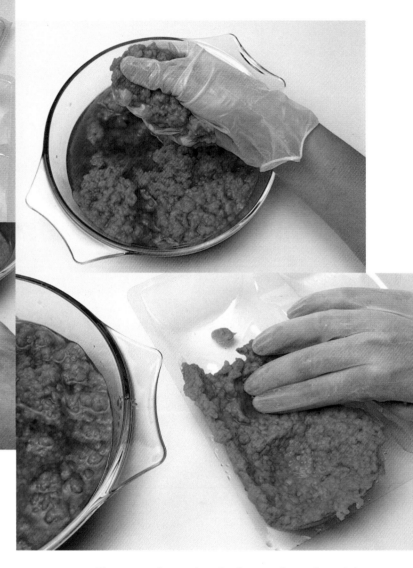

This pack has two shapes to provide an interesting
contrast between the two methods. Cover the inside
surfaces of both shapes (molds) with petroleum jelly
to act as a releasing agent.

2

Place a small quantity of pulp onto the surface of the
mold and gently tap it into place with your fingers.
As you do so, water will rise to the surface. Maintain
an even thickness of pulp over the whole surface,
especially over raised areas of the mold. It is easy to
fill the cavities with pulp and to forget to add enough
to the raised areas.

4

5 Cover the entire surface of the mold with pulp, making sure that it is well compacted and of an even thickness. Use a sponge to soak up the excess water that has risen to the surface and at the same time push the pulp down against the surface of the mold.

6 To take the cast from the other side of the blister pack, take a sheet of wet paper that has been pressed but not dried, and tear it into small strips about ½ x 1 inch long.

7 Lay a strip down on the surface of the mold, and push it well into the surface so as not to trap any air. Partially cover this with another strip. Holding a brush vertically, gently tap the strip down onto the mold surface and part of the adjacent piece of paper. Continue to overlap the strips in this way to cover the whole surface. Repeat this technique to place another layer of paper on top. Use the sponge to absorb any water on the surface, then let the paper dry.

8 Once the paper has dried, remove it gently from the mold to reveal a copy of the mold's interior, while at the same time showing the different surface textures of the two casting methods.

MAKING YOUR OWN MOLD IS EXCITING AND FULFILLING, ALLOWING TOTAL CONTROL OVER THE

FORM AND COMPOSITION OF THE CAST. BY MAKING A SIMPLE MOLD USING MODELING CLAY, A MATERIAL

WHICH IS EASILY AVAILABLE, IT IS POSSIBLE TO CREATE AMBITIOUS PAPER CASTS. THE PROJECT SHOWS THE TECHNIQUE OF MAKING

technique

embedding in clay

a press mold, which provides a very easy and direct way of obtaining a negative impression of an original. The advantage of the press mold is that it allows a lot of experimentation. If the impression of an object is unsuccessful, the clay can be rolled flat and another object pressed in, until a satisfactory result is obtained. Pressing an object into the clay too far can make removal difficult and increase the likelihood of damage. Once a cast has been taken from the mold, it can be used again for another cast; conversely, the clay can be rolled flat to make a different mold.

materials

modeling clay

wooden shape

shell

petroleum jelly

pulp

equipment

roller

cloth

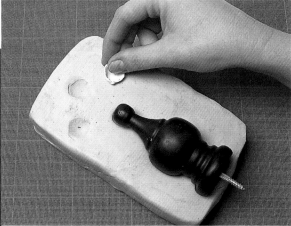

Using the size of the shapes to be impressed as a guide, roll out a bed of modeling clay 2 inches thick and large enough to accommodate them. Don't forget to allow for a margin around the edge.

Position the wooden finial on the clay and push down as hard as possible to embed it. Push the shell into the clay, lifting it out carefully to leave its impression. Continue around the top of the finial to complete the composition.

2

3 Carefully remove the finial to reveal a negative impression of its shape. Apply petroleum jelly to the entire surface, taking care not to damage the impressions in the clay.

4 Take a small quantity of pulp, gently squeeze out a small amount of water, and apply it onto the surface of the mold, pushing it into place. Remember to maintain an even thickness of pulp over the whole surface, especially raised areas of the mold, and to compact the pulp together well.

5 Use a cloth to soak up excess water. Have a bowl nearby in which to squeeze the water from the cloth. At the same time, compact the pulp with the cloth to make sure the pulp is bonded together well.

6 When the cast is dry, lift off the clay. This technique is highly efficient in picking up the fine detail of the shell. The edge has been left rough to complement the interesting surface texture of the whole cast.

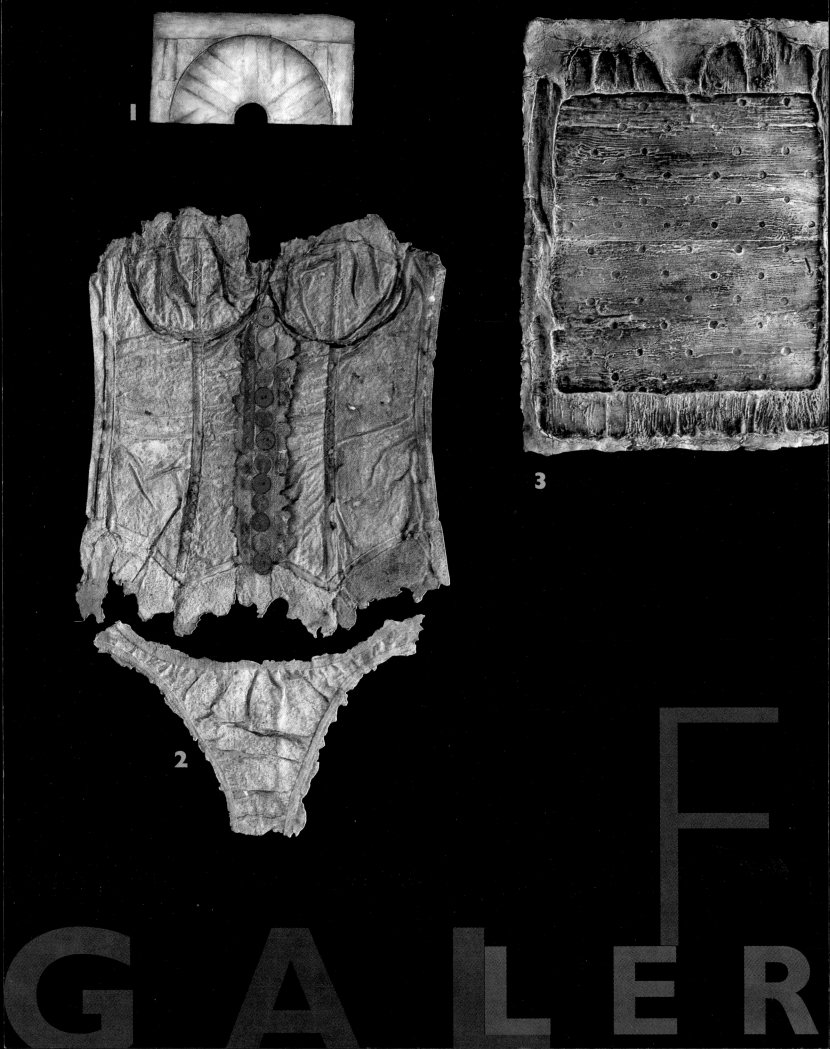

1.

3

2

GALLER F

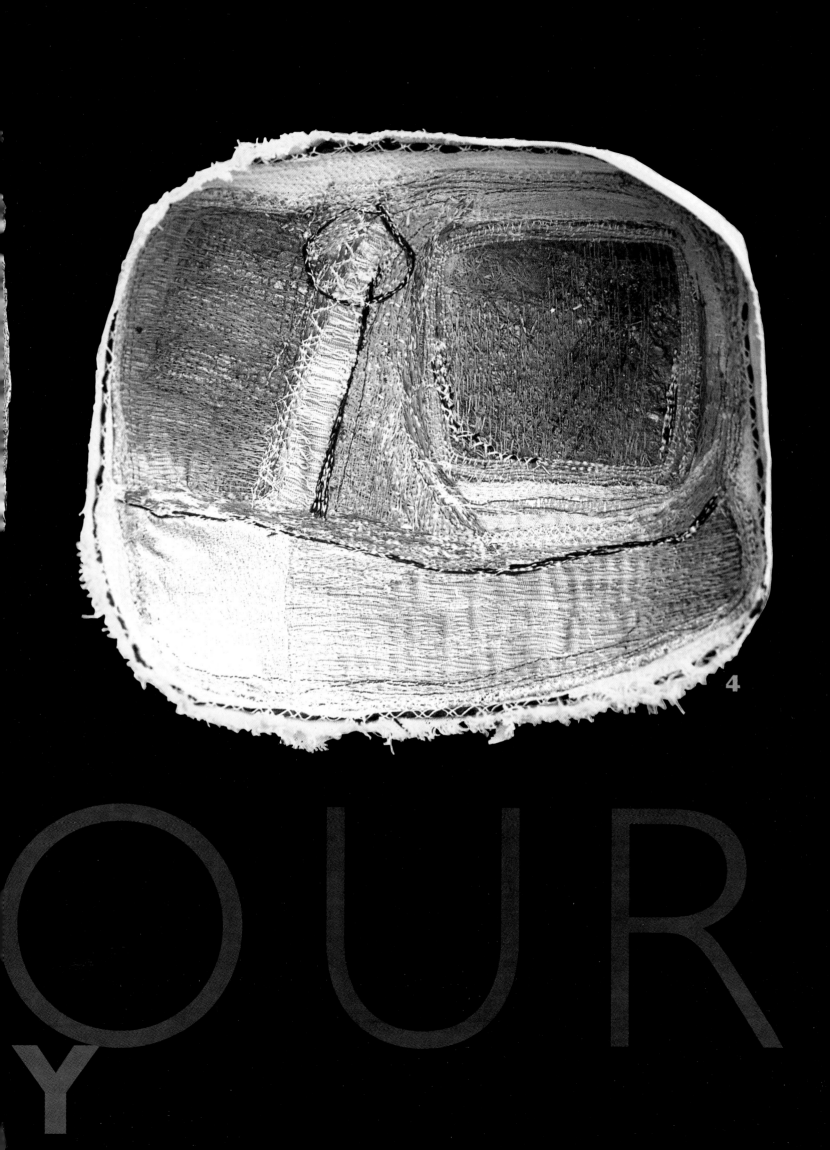

4

YOUR

THIS ATTRACTIVE BOWL WOULD MAKE A WELCOME ADDITION TO ANY TABLE TOP OR SHELF. IT COULD STAND ALONE OR BE USEFUL FOR DISPLAYING DRIED FLOWERS OR OTHER INTERESTING OBJECTS. ITEMS EMBEDDED INTO THE WET PULP BECOME AN INTEGRAL PART OF THE BOWL, MAKING IT UNNECESSARY TO PROVIDE ANY

p r o j e c t

thread bowl

additional decoration. When adding the colored pulps to the mold, it is essential that they are compressed tightly together. Taking this cast from the inside of a bowl provides a vivid example of the shrinkage of paper as it dries, producing a smaller cast. When the bowl has dried, it will be extremely resilient, and strong and tough. Any interestingly shaped bowl can be used, but always remember to take the cast from the inside.

m a t e r i a l s

petroleum jelly

plastic wrap

colored threads

colored pulps

pieces of felt

e q u i p m e n t

rubber gloves

glass bowl

cloth

t e c h n i q u e s

casting

embedding

GALLERY FOUR *pages 94-95*

1 Carol Farrow, "Moulin monle", cast of half a mill wheel, with wooden surround.

2 Elizabeth Couzins, handmade paper using silks, threads, and found materials.

3 Carol Farrow, "shutter," cast.

4 Amanda Goode, paper and fabric bowl.

1 Line the inside of the bowl with plastic wrap, pushing it flat against the surface. Smear petroleum jelly over the plastic wrap to hold the threads and felt in place.

2 Lay threads across the inside of the bowl, pushing them into the petroleum jelly and let the ends of the thread hang over the sides. Add pieces of felt to the sides, pushing them into the jelly.

3 Place pink pulp in the center of the bowl and add green pulp to either side of it. Make sure the pulps are of an even thickness and well compacted together.

4 After the two small areas of blue pulp have been added, use a cloth to soak up the excess water, at the same time tapping the pulp down. Notice that the pulp does not go up to the edge of the bowl, but stops short to produce an interesting uneven edge. Lay more threads on the surface of the pulp and tap them down with a cloth. Once dry the cast can be removed quite easily.

FOR MORE DETAILED AND SOPHISTICATED SHAPES, MOLDS MADE SPECIFICALLY FOR TAKING PAPER

CASTS CAN BE PURCHASED FROM SPECIALIST CRAFT SHOPS. THEIR ADVANTAGE IS THAT NO SPECIAL SKILL IS

NEEDED TO PRODUCE THE REALISTIC SHAPES FROM WHICH THE MOLD HAS BEEN MADE. ONCE THE SHAPES HAVE BEEN CAST, THE

project

ready-made molds

creativity comes in deciding how to use

them. They are ideal to use as embellish-

ments for presents, decorative boxes, or

greeting cards. They can be turned into fridge magnets, used to make a mobile, and make excellent gifts for children.

materials

colored pulps

petroleum jelly

equipment

ready-made mold

toothbrush

cloth

techniques

casting

1 Cover the inside of the mold with petroleum jelly. Use the toothbrush to make sure it gets into all the details on the surface.

3 When the pink pulp has been put into the shell, use a cloth and tap the pulp down to soak up the excess water lying on the surface. Squeeze water from the cloth into the bowl and continue mopping up. Leave the casts to dry.

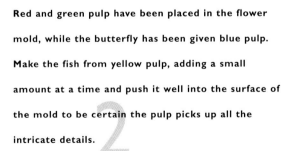

2 Red and green pulp have been placed in the flower mold, while the butterfly has been given blue pulp. Make the fish from yellow pulp, adding a small amount at a time and push it well into the surface of the mold to be certain the pulp picks up all the intricate details.

4 When the casts are dry, go around their edges carefully with a knife to tease them gently out of the mold. As you can see, the fish cast is an exact replica of the surface of the mold.

THE KITCHEN IS A GREAT SOURCE OF MOLDS SUITABLE TO USE FOR CASTING PAPER, SPECIFICALLY THOSE USED FOR GELATINS, MOUSSES, AND CAKES. WITH A BIT OF IMAGINATION, THE CASTS FROM THESE MOLDS CAN BE USED TO MAKE A VERY INTERESTING WALL HANGING. THIS PROJECT, WHICH DEVELOPS TECHNIQUES ALREADY

wall fish relief

covered in the book, uses a fish mold from which a cast is taken. The cast is mounted on a backing of sheets that have been laminated in a sea-colored net configuration to complement the fish. In the resulting wall hanging, all the elements go together to produce a marine composition.

materials

formed sheets: turquoise, gray, pink, blue, and white

petroleum jelly

techniques

sheetforming

casting

laminating

equipment

ruler

roller

stones

plastic wrap

fish-shaped mold

brush

sponge

1 Position a ruler on top of a formed and pressed sheet of gray paper, ½ inch away from its long edge. Carefully lift one end of the resulting strip of paper and bring it back slowly, tearing the paper against the edge of the ruler. Reposition the ruler and continue to tear strips from the rest of the sheet. The strips can be used whatever their length, so don't worry if they break during tearing.

2 Lay two wet sheets of green paper side by side with their long sides overlapping. Lay a felt over this seam and roll it flat to make a larger sheet. Lay the strips of gray paper diagonally across the sheet.

3 Position more diagonal strips going in the opposite direction to create a mesh effect. The diagonal strips have been made from varying lengths of paper and continue over the edges of the sheet.

4 Lay two sheets of green paper on top, overlapping as before, lining up their outside edges with those of the bottom sheet. When they are positioned, lay a felt on top and roll in all directions to make sure that the paper is compacted together thoroughly.

5 Although the sheet is wet, it will be strong enough to perform this maneuver. Carefully lift the laminated sheet and lay it on top of an arrangement of stones. Each stone has been wrapped in plastic wrap to prevent the paper from sticking as it dries.

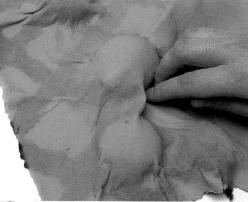

6 Push the sheet down gently so that it molds itself around the stones to create a wavy surface. Do not push the paper under the stones as they can become trapped in the paper when it has dried.

7 Leave the sheet to dry so that it can easily be lifted off the stones. The sheet is strong and self-supporting, with an attractive firm, wavy surface.

8 Apply petroleum jelly to the inside of the fish mold, working it well into the detailed areas of the mold.

9

Tear a blue sheet of paper into small strips
½ x 1 inch and place them in the fin and tail areas,
overlapping one another. Use the brush to push them
together, making sure that all the detail from the
mold is picked up by the paper. Mop up any excess
water with a sponge.

11

Finally add two layers of strips of white paper. These
additional layers will strengthen the cast when the
paper is dry, and using a different color paper helps
to indicate whether the entire surface of the mold
has been covered.

Continue covering the rest of the surface of the mold
with strips of pink paper. Continue laying the paper
down up to the edge of the mold; don't be concerned
with getting an even edge. Use the brush to tap it
down well, mopping up the excess water with a
sponge. You would need to make two layers of the
colored sheets.

10

When the paper has dried, go gently around the
edges with a knife and lift the edges away from the
mold. Turn the mold over and take it off the cast.

12

IDENTICAL SECTIONS, JOINED BY TEARING THE PAPER AND SLOTTING EACH UNIT TOGETHER, ARE REPEATED TO FORM THIS FREESTANDING SCULPTURAL COMPOSITION. THIS METHOD RESPECTS THE ATTRIBUTES OF THE MATERIAL ITSELF, RATHER THAN RELYING ON ARTIFICIAL MEANS TO HOLD EACH UNIT TOGETHER. THIS THEME OF SIMPLICITY

table sculpture

is continued when casting the shape of each unit, which is obtained by draping a sheet embedded with wooden skewers over a cylinder. The sticks give the cast shape the strength and extra stability necessary for a freestanding sculpture.

materials

wooden skewers

formed sheets

equipment

roller

plastic drainpipe

techniques

sheetforming

laminating

embedding

casting

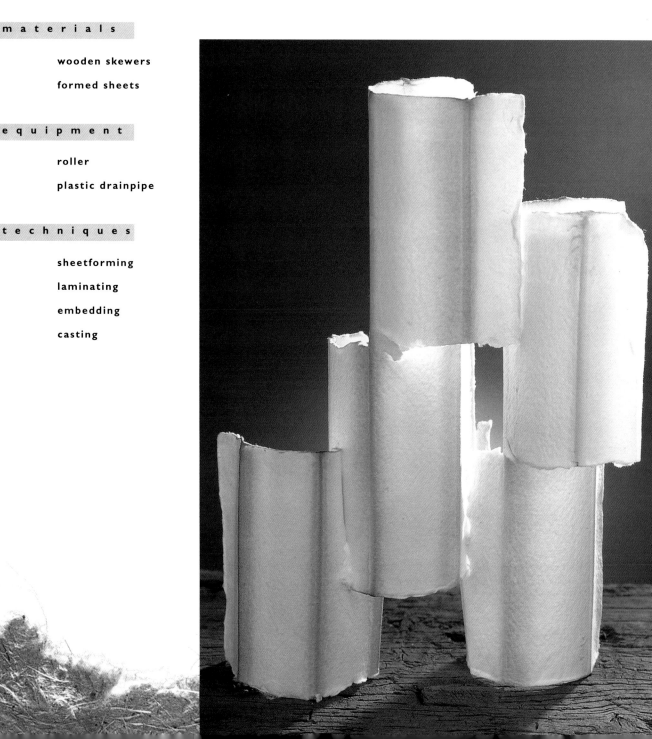

1

On top of a couched sheet, position three wooden skewers lengthwise, equidistant from one another.

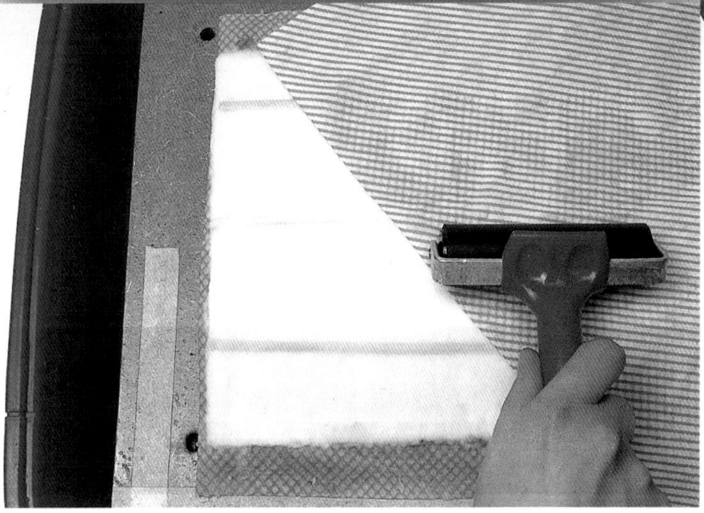

2

Laminate a sheet directly on top by laying a felt on top and going over the whole sheet with the roller pressed hard. Make sure to roll right up to the skewers to expel any pockets of trapped air.

3

Lift the embedded sheet carefully and lay it lengthwise on a length of plastic drainpipe with the central skewer positioned centrally on top. Let the paper hang down evenly on each side, lay a felt on top, and carefully roll over the paper.

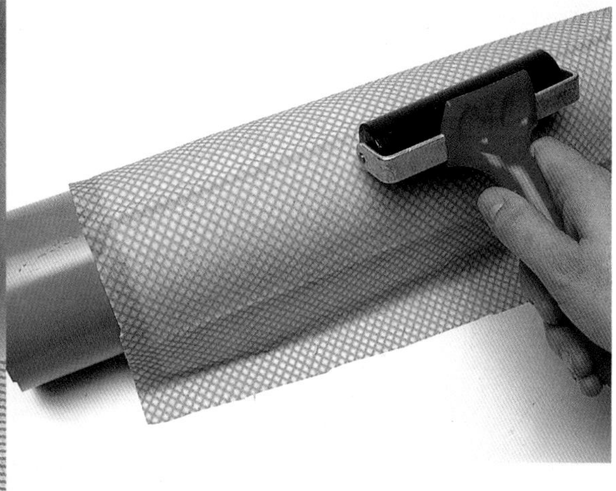

4

Peel back the felt to reveal the wet sheet, which is now firmly positioned on the pipe. Leave the sheet on the pipe to dry. Lay down and position two more sheets on the remaining length of pipe.

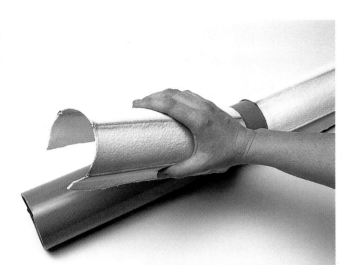

5 When the sheets are dry, carefully lift the three casts from the pipe. The embedded skewers make each cast extremely rigid.

6 Divide each of the casts in half widthwise. Tear the paper up to the wooden skewer, cut this with scissors, and continue tearing the paper. Repeat across each cast to obtain equal-sized shapes.

7 Although there are six casts, five is a better number to make a composition with. Move them around, altering their position in relation to one another until you find an interesting arrangement. These units can be slotted together, providing a strong and stable structure, by tearing the paper vertically halfway down.

THE FRAME USED HERE IS INTEGRATED INTO THE CONSTRUCTION OF THE LAMPSHADE ITSELF. ITS DESIGN FULLY EXPLOITS THE QUALITY OF LIGHT SHINING THROUGH PAPER EMBEDDED WITH CONFETTI, WHICH REVEALS A MULTITUDE OF INTERESTING SHAPES WHEN IT IS HELD UP TO A LIGHT. THE SHEETS OF EMBEDDED PAPER NEED TO BE

project

corrugated lampshade

made as thick as possible to provide maximum rigidity when they are dry, so have plenty of pulp in the tub when forming them. The sheets of paper are positioned on top of corrugated plastic and left to dry to produce casts of corrugated paper. This hanging lampshade needs to be sprayed with a flameproofer specifically for use on paper, and should not be used with a bulb of more than 60 watts.

materials

4 sheets of embedded paper

4 lengths of 3-sided wooden beading

4 lengths coat-hanger wire

flameproofer spray

white glue

equipment

corrugated plastic large enough to fold the sheets of paper separately

roller

pliers

knife

techniques

sheetforming

embedding

casting

Couch and lightly press an embedded sheet of paper, then position it over the corrugated plastic while it is still on the felt. Gently push it over and down into the corrugations. To make sure the sheet is positioned squarely on the plastic, line its long sides up with the corrugated ridges.

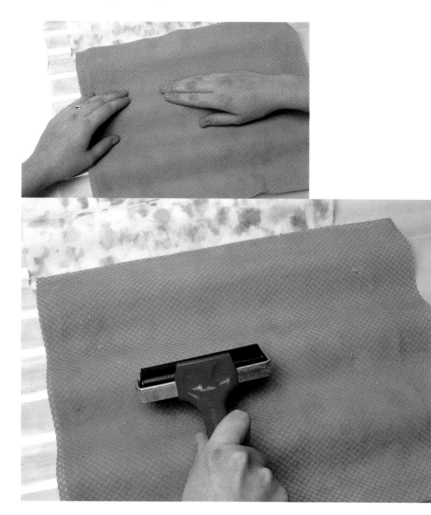

After rolling, lift the felt carefully and leave the sheets of paper to dry on the plastic. When they are dry, use a knife blade to release the edges of the paper from the plastic, then gently lift the corrugated cast away from the plastic.

3

When the sheet is in position, roll over the felt, pushing the paper well down onto the surface of the plastic and pushing out any air that may be trapped. Repeat this process, laying down three more sheets.

2

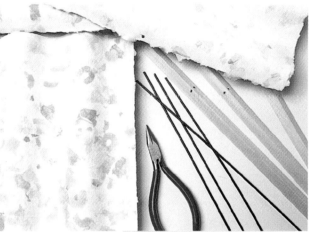

4 Cut the four lengths of triangular wood to the length of the paper. In each piece of wood drill two holes, side by side, to half the depth of the wood and 4 inches from one end. The diameter of the hole must be the same as that of the wire.

5 Glue two sheets of corrugated paper together along the long edges. Apply glue along two sides of the wood and place it on top of this seam. Fold each side of the paper up against the wood to create one corner. Repeat with the other two sheets. Stand the two corner units upright and join them in the same way to form a box shape, making sure the drilled holes are at the same end.

6 Stand the lampshade up and square it to produce the box shape shown, and let the glue dry. Measure diagonally from hole to hole and cut each wire slightly longer. Bend each wire slightly and insert each end into the opposite drilled holes. In addition to adding extra stability to the lampshade, the wire cross frame also provides a hanging device for the lampshade. The cord can be threaded through the central square formed by the diagonal wires.

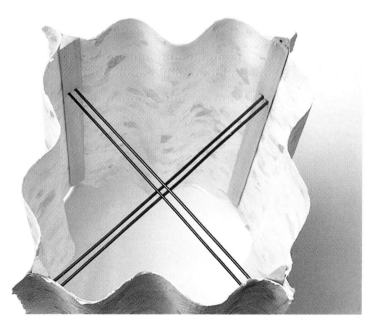

5

dry paper

Dry paper is, of course, the paper we are most familiar with. Creating books, cards and stationery with handmade paper, however, makes them that much more characterful and attractive. Books have a timeless quality that means they will never lose their appeal, and the simpler they are, the more beautiful. In China, the earliest books consisted of thin slips of bamboo or wood, which were joined by cords. These were replaced by silk scrolls, which were in turn replaced by paper scrolls. The nature of the scroll meant that a long one proved impractical to use – all that unrolling. So scrolls were folded, like a concertina with covers attached to the first and last sections. Later, single sheets of text began to be folded face inwards, stacked (for binding) and then pasted along the folded edges. The book as we know it had been born. Today, papermakers still make books by hand, from handmade paper, creating wonderful collections with varied shapes, sizes, bindings and decorative covers. Here, we look at single section books and Japanese-style books. The latter were developed during the 11th and 12th centuries in Japan. Consisting of multi-section stitched binding (*retchoso*), they were made by stacking several sheets of paper, folding them in half to form a section, then stitching sections together through the central folds. In our project we use a variety of plant papers to give a subtle, organic effect. Stationery, cards and envelopes made from handmade paper have a similar appeal to books made from the same.

THIS IS THE SIMPLEST CONVENTIONAL WESTERN METHOD OF BINDING A BOOK. BECAUSE THE PURPOSE OF THE BINDING IS PURELY FUNCTIONAL AND NOT DECORATIVE AT ALL, IT IS HIDDEN. BEFORE YOU START ANY FOLDING, MAKE SURE YOU HAVE ESTABLISHED THE GRAIN DIRECTION — IN OTHER WORDS, THE WAY THE FIBERS ARE

pointing. It will then be much easier to fold along the grain and ensure a smooth, professional looking finish.

project

single section book

materials

marbled paper
red paper
white paper
heavy thread

equipment

bone folder
steel ruler
needle
cutting mat

techniques

tearing
creasing
folding
sewing

1 Use the white paper for the pages, the red sheet for the interleaf, and the marbled paper for the cover.

2 Mark five sheets, each 8 x 10 inches, on the large sheet of white paper. Fold and crease along the marked line, then insert a steel ruler into the fold and pull it through to tear the paper.

3 Tear one sheet of red paper 8 x 10 inches and place the five white sheets neatly on top. Fold the stack in half and use the bone folder to press down firmly, bringing it along the fold toward you to crease the sheets. Insert the pages inside a folded and creased sheet of marbled paper which is slightly larger than the pages.

4 Open the book and mark five equidistant points along the central spine. Make a hole with an awl at each of these points. Take the threaded needle through the center hole and pull it out, then take it through the next and into the top hole. Make sure that the thread is kept taut at all times.

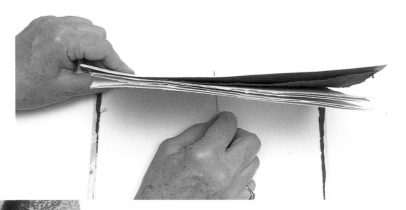

5 Come out of the top hole and return the thread through the next hole. Continue down the spine. When returning through a hole, be very careful not to break through the thread already there. Continue along the spine until you reach the other end and return to the middle hole.

6 To finish, come back up through the middle hole, pull the thread tight, and tie it in a nice tight knot. Fold the book over and place it in a press or under a pile of heavy books for 24 hours to reinforce the fold.

THE BINDING USED HERE IS NOT HIDDEN, BUT BECOMES AN INTRINSIC PART OF THE DECORATIVE DESIGN OF THE BOOK. A BASIC STITCHING TECHNIQUE IS SHOWN, BUT VARIATIONS ON THIS STITCH CAN ALSO BE USED. WHATEVER STITCH IS CHOSEN, IT RESULTS IN A VERY INTERESTING DECORATIVE BINDING. A BOOK SUCH AS THIS WOULD

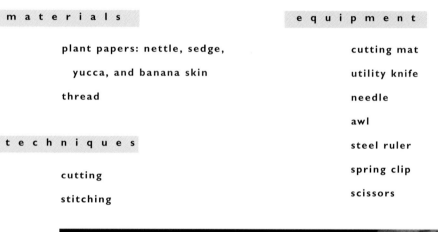

project

japanese-style book

make an excellent gift. Use it as an album, and fill it with photographs, or as a scrapbook. The thick nature of the pages would make it perfect for either of these things.

materials

> plant papers: nettle, sedge, yucca, and banana skin
>
> thread

techniques

> cutting
>
> stitching

equipment

> cutting mat
>
> utility knife
>
> needle
>
> awl
>
> steel ruler
>
> spring clip
>
> scissors

Banana skin, sedge, nettle, and yucca plant papers have each been torn to 8 x 10 inches. Two sheets of banana skin paper will be the cover while the other papers will alternate inside as the pages of the book.

Stack the papers with the banana skin at the top and bottom and hold them in place with a spring clip. Place a piece of scrap paper under the clip on each side to protect the banana skin paper. Using the steel ruler and utility knife, cut the opposite end of the pile to produce a clean, straight edge.

2

3 Very faintly draw a line ⅝ inch in from this edge, and then, working up from one end, clearly mark four points at ½ inch, 1¾ inch, 1¾ inch, and ½ inch spacings. Use an awl in order to pierce through the stack of paper at each of these points.

4 Starting at the second hole from the top, come up from the bottom and down through the next hole. Pull the thread through and bring it around the edge and back down through the same hole.

5 To make the corner stitching, lay the thread at a right-angle to the spine thread. Keeping it taut, bring the thread up through the top hole and pull tight.

6 Continue back down the spine, going through the holes again. Be careful not to tear the thread already in the holes. When you reach the other corner, make a right-angled stitch as in **Step 5.**

Book: Purples/browns series. **Pat Hodson.**

7 Turn the book over and then, making sure to hold the other end of thread taut, take the needle under the stitching on the spine.

8 Bring the needle through, keeping both ends of the thread taut. Tie a knot and cut off the loose ends.

IN THIS PROJECT YOU WILL BE BRIGHTENING UP AN OTHERWISE ORDINARY TABLE. THE TABLE IN THIS EXAMPLE HAPPENS TO BE NEW, BUT THE SAME METHOD CAN BE USED FOR ANY TABLE. DON'T WORRY ABOUT BEING TOO EXACT WHEN TEARING OUT THESE PAPER SHAPES. THE TEXTURED PAPERS AND THEIR ROUGH EDGES ARE ONE OF THE

table top collage

charms of this collage. Once glass has been laid on top, it will flatter and enhance your collage and provide a practical surface.

materials

small table

glass cut with smoothed
  edges to the same size
  as the table top
white glue
colored handmade
  decorative paper

equipment

steel ruler

pencil

roller

brush

techniques

tearing

gluing

I

This project uses eight brightly colored handmade papers from India. Because it is thin, it is sold as gift wrap, and is ideally suited to collage work. You can also use your own handmade papers.

3

Tear paper to correspond to each of the defined areas on the table top. To get the correct shape, lay the paper over the area and tear it carefully so the torn edge corresponds to the drawn line. Make sure the paper is held firmly in position on the table top when tearing.

On the table top, mark a design based on the dimensions of the table. Draw diagonal lines from corner to corner. Where the lines cross in the middle, draw two concentric circles around the center point. Divide each corner area diagonally in two.

2

Apply glue to the underside of a piece of torn paper and position it on the table using the pencil marks as a guide. Making sure there are no air bubbles, roll the whole piece flat.

4

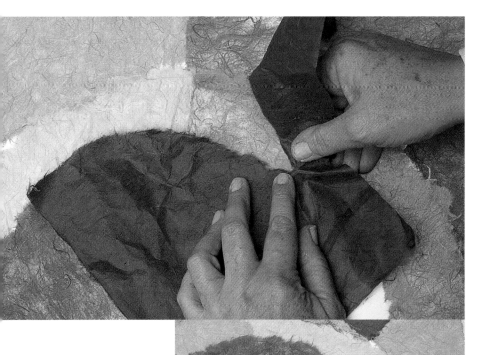

5

Working toward the center, tear the circular edge of this segment.

6

To tear the two straight edges, place the ruler along the drawn line and pull the paper up and back.

7

Having stuck all the paper shapes down, for a final compositional touch, add a circle of paper to the center and roll flat. Leave for 24 hours to let glue dry. Place a piece of glass the same size on top.

THIS IS A VERY SIMPLE AND STRAIGHTFORWARD PROJECT WHICH PRODUCES A VERY ATTRACTIVE CARD. THERE ARE ENDLESS VARIATIONS — THE ONLY LIMIT IS YOUR OWN IMAGINATION. YOU CAN EITHER USE PAPERS THAT YOU HAVE MADE, OR CHOOSE FROM THE VAST ARRAY OF HANDMADE PAPERS THAT CAN BE BOUGHT. HERE WE USE tea paper and onion paper to contrast against the main body of the card. Although the design is undoubtedly simple, the different textures and colors of the handmade papers provide an attractive card. You can build on this simple idea to create cards that can be used for any occasion.

project

greeting card and envelope

materials

handmade papers

white glue

equipment

steel ruler

utility knife

hole punch

bone folder

small brush

pencil

double-sided tape

techniques

tearing

cutting

scoring

folding

gluing

1 Measure and mark a rectangle 8 x 10 inches on a sheet of paper. Score the paper by placing a ruler on the lines, then pull a needle along the straight edge using just enough pressure to break the surface of the paper.

2 Holding the paper firmly, tear down one side of the scored line.

3 Fold this sheet in half and crease it with the bone folder to compact and consolidate the fold.

4 Position a sheet of tea paper squarely in the center of the card and draw around it.

5 Coat one side of the tea paper evenly with glue, then mount it on the front of the card using the drawn lines as a guide. Push it down firmly into place. Now make an envelope to fit the card. First mark the envelope shape, lay the open card on a sheet of heavy paper, and draw around it, leaving a 1-inch margin on each side and at the top.

6 Shade the waste areas and cut them away along the marked lines with a utility knife.

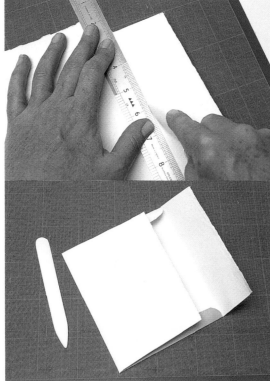

7 To make it easier to fold the paper, score it by holding a steel ruler on each of the drawn lines and using slight pressure pull the bone folder toward you.

8 The edges of the top flap have been cut inward to form a wedge shape. Glue the two sides in place and stick a length of double-sided tape along the top flap to be used when the envelope is sealed.

9 The completed card with an onion paper added fits snugly into the envelope. This method of making an envelope can be adapted for any shape card.

THIS PROJECT DEMONSTRATES THE EFFECTIVENESS THAT CAN BE ACHIEVED BY USING DIFFERENT

WEIGHTS OF PAPER. THE CONTRAST IS ALL THE GREATER BECAUSE OF THE COARSER TEXTURE OF THE MAIN

BODY OF THE CARD. ONCE COMPLETE, THE LIGHT SHINING THROUGH THE RED "HEART" MAKES FOR A VERY DISTINCTIVE VALENTINE'S

valentine's card

greeting. Again, the technique is simple, but be

sure to make the card carefully. Attention to

such details as using a roller to get rid of

bubbles will ensure a professional, as opposed to an obviously home-made, result.

1

After tearing one sheet 8 x 10 inches and a thin red
sheet 7½ x 9½ inches, fold each one in half and
crease. **On the thicker paper draw a heart shape in
the top right-hand corner, then go around the drawn
line, piercing it with a needle.**

2

These pierced holes will allow you to tear the heart
shape carefully away from the paper.

3

Apply glue to the inside of the front of the card.
Stick on the red paper, using a roller to get rid of any
air bubbles that may have accumulated.

YOU CAN USE THE MARBLED PAPER YOU HAVE ALREADY MADE TO MAKE AN ATTRACTIVE PICTURE FRAME THAT WILL PROVIDE THE PERFECT COMPLEMENT FOR YOUR FAVORITE PHOTOGRAPH. ALTHOUGH THE TECHNIQUES USED ARE SIMPLE, IT IS ESSENTIAL TO CUT, FOLD, GLUE, AND MEASURE VERY CAREFULLY TO ACHIEVE A PROFESSIONAL-looking finish. Any ill-positioning and rough edges will detract from the decorative attractiveness of the paper.

project

photograph frame

materials

mat board

marbled paper

white glue

equipment

ruler

pencil

utility knife

steel ruler

roller

cutting mat

techniques

cutting

gluing

1

Measure and mark two rectangles, each 10 x 12 inches, on the mat board. Within one of these, mark another rectangle 5 x 7 inches, and 2½ inches in from each side and 2½ inches down from the top. The third shape measures 3 x 8 inches for the stand. Using a utility knife and working on the cutting mat, cut these shapes from the board. The stand shape needs to be scored 1½ inches down from the top.

3

Cut out the marbled paper, 11¾ x 13½ inches for the front, and 10 x 12 inches for the back. The stand needs two pieces 5 x 9 inches and 3 x 8 inches. Place the front board section on the back of the marbled paper, draw around the outside and the window, then draw a ⅜-inch margin inside the lines of the window.

Cut a wedge shape for the stand, 8 inches long and 3 inches at one end, tapering to 2 inches at the other end. Measure 1½ inches down from the tapered end, then score along this line and fold over slightly.

2

Brush glue onto the board, and lay the marbled paper on top, correctly positioning it, and roll it flat. Fold back the flange running around the window and glue it onto the back of the board.

4

5 Add two strips of thinner cardboard, ½ x 11 inches along each of the sides and a piece ½ x 9 inches along the bottom edge as spacers. Cut the top of the marbled paper and fold over the top edge; glue.

6 Coat the three strips of spacer card with glue, position and glue in place on the backing board.

7 Fold over and glue the flanges of the front piece of marbled paper. Coat the back with glue and position a piece of marbled paper the same size on it. Cover the stand with marbled paper in the same way and glue it along the hinge to the back of the board.

FOR THIS PROJECT YOU NEED A MUCH STRONGER HANDMADE PAPER, SUCH AS HEAVY INDIAN KHADI WHICH COMES IN A VARIETY OF COLORS. CHOOSE SHADES THAT WILL COMPLEMENT EACH OTHER. HERE WE HAVE RICH SHADES OF BROWN TO GIVE AN EARTHY, ELEMENTAL LOOK. WEAVING IS A VERY EASY TECHNIQUE, AND

project

wastepaper basket

increases the strength of the basket (wastepaper baskets can take quite a battering!) Not only useful but attractive too, this basket provides a nice irony in the fact that you are using paper to create an object whose function is to dispose of waste paper.

materials

> 3 different-colored khadi
> papers
> thin handmade decorative
> paper
> white glue

equipment

> steel ruler

techniques

> tearing
> weaving

1

To begin you need a supply of paper in three

contrasting colors torn into strips of the same width.

2

To produce these strips lay the ruler along the length
of the paper, lining up the left-hand edge with the
edge of the sheet. Holding the ruler firmly, lift the
paper. By pulling it slowly, the paper will tear along
the straight edge. The strip you need will be under
the ruler, so using this method you can produce as
many strips of the same size as you need.

3

Place five strips of paper side by side making sure
they are butted up to one another. Start to weave
five more strips over and under these at right-angles.

4 You have now woven ten strips at right-angles to one another to produce the bottom of the basket.

5 Bring the four sides, each containing five strips, up at right-angles to the bottom. Working one strip at a time, place your finger along the fold and firmly bring the strip up vertically, then bend it back and crease to reinforce the fold.

6 You may find starting the weaving process around the sides a bit tricky, and it may be helpful to prop up the side strips to get you going. Start weaving in and out with more strips as you did for the bottom.

7

Once you have put in two strips all around, you will find that the basket will become self-supporting. Continue weaving; pull the strips tight, making sure they are packed close together.

8

To finish the top, cut the upright strips square and level, and glue them to the last horizontal strip. To hide the edge, roll up a strip of light handmade tissue paper in a slightly haphazard fashion and glue it to the top of the basket. Once the glue is dry, coat the inside of the basket with diluted glue, which will dry colorless and add strength to the paper.

THIS SHADE IS THE PERFECT REMEDY FOR AN UNSIGHTLY VIEW, OR IT CAN BE APPLIED TO ANY

GLAZED AREA TO ADD ANOTHER DIMENSION TO INTERIOR DECORATION. LIGHTWEIGHT EMBEDDED PAPERS ARE

USED TO MAKE A COLLAGE IN WHICH SHAPES ARE STUCK ONTO A THIN BACKING SHEET OF WHITE EMBEDDED PAPER FITTED TO

project

window shade

cover the glazed areas of each window. To capitalize on the

exciting texture of a torn edge, these shapes have been torn

rather than cut out from each sheet. When you select the

sheets to use, make sure the colors and textures complement each other and the decor near the window.

materials

selection of thin handmade
decorative papers
white glue

equipment

utility knife
pencil
brush

techniques

creasing
folding
tearing
gluing

Incised pattern, Series No. 2.
Etched monoprint on
handmade emmer paper with
muslin. **Marie Wright.**

1 The following types of paper (from right to left) have been used to make this window shade: Bark and Tobacco, Mulberry, Mixed Leaves, Rag and Banana, Japanese fiber, and Bamboo and Japanese fiber. The two white sheets are used to make the backing sheets onto which the torn shapes will be stuck.

2 Make a template of the leaf shape. Fold a sheet of paper in half, and draw half of the leaf shape on it. Cut out the shape from the paper and unfold it to reveal a symmetrical shape. Place the template on the bamboo leaf paper and draw around it. Carefully tear out this shape using the pencil marks as a guide. This paper is embedded with leaves that may be a hindrance to easy tearing; when you reach a leaf, tear more slowly.

To prepare each of the backing sheets hold the paper up to the window, creasing the paper where the glass joins the wooden frame. (In this case the window divides into two, the top section being arched, the bottom rectangular.) Fold the corners as shown and glue down with the other folds and creases.

Lay the leaf shape on top of another type of paper (in this case mulberry) and tear the paper following the profile of the side and lower sections of the leaf. Tear the rest of the sheet to fit the frame, leaving a margin at the sides but none at the bottom. Then take another sheet of paper and do the same with the top half of the leaf, again leaving a margin around the frame edge. Glue both background shapes onto the backing sheet and then glue on the leaf shape. The upper section is now complete.

5 For the lower section of the shade, tear a central rectangle using the rest of the bamboo paper. Tear shapes from the remaining sheets of paper to fit around the rectangle, leaving a margin around the frame edge, deeper at the bottom. Do not leave a margin at the top where the two shade sections join. Glue the reverse of each piece of paper and place on the backing sheet. Go over with a roller to make sure it is pressed well down.

6 Both the bottom and top sections have been made and should be left for the glue to dry. Notice that the only straight edges are those that delineate the border of the shade.

1

3

2

GALLER

4

FIVE
Y

glossary

Alkali A caustic substance used in COOKING plant fibers to remove gums, waxes, starch, and other non-cellulose materials.

Bamboo Grass fiber from the stem of the bamboo plant *(Phyllostachys aurea)*, used for papermaking by the Chinese.

Beating Separating and macerating fibers into pulp for sheet formation, done by hand or mechanical beater. See HOLLANDER.

Bleaching A process used to purify and whiten pulp.

Bleed The spread or feathering of ink or color within a sheet of paper.

Bonding The capacity of cellulose fibers to adhere and combine. BEATING and drying promote bonding.

Calcium Carbonate Used primarily to promote longevity in paper. In larger amounts it acts as a filler to retard shrinkage in paper casting, and in sheetforming to improve opacity and whiteness.

Cellulose The chief component of plant tissue, which provides the basic substance for paper manufacture in the form of FIBER.

Collage The pasting together of various materials to create an image.

Cooking The treatment of raw fibers to promote separation, remove contaminants, and dissolve unwanted plant material, usually achieved by heating in an alkaline solution.

Cotton The soft white filaments attached to the seeds of the cotton plant, one of the main fibers in Western hand papermaking.

Cotton linters The pulp produced from the shorter seed hairs of the cotton plant which has been cooked, bleached, beaten, and made into compressed sheets.

Cotton rag A long-fibered pulp made from new rag cuttings.

Couching Transferring a freshly made sheet of paper from the MOLD surface onto a dampened FELT (Western papermaking) or directly onto the previously couched sheet (Japanese papermaking).

Cutting An optional function of beating used to shorten the fibers.

Deckle The removable frame which fits on and around the MOLD cover to contain the pulp and determine the size of the sheet.

Deckle edge The distinctive, slightly ragged edge of a sheet of handmade paper, created by a small amount of pulp seeping under the DECKLE during formation.

Dyes Soluble coloring agents which penetrate the structure of a fiber and become attached to it.

Embedding Incorporating materials in a sheet of paper, so that the fibers hold the embedded material in place.

Embossing The creation of a raised or

depressed surface design in a sheet of paper.

Felt The woven woolen blanket onto which a newly formed sheet of paper is transferred, or COUCHED, in traditional Western hand papermaking.

Fiber The threadlike structure in plant tissue from which papermaking pulp is made.

Finish The surface qualities of a sheet of paper.

Flax A bast fiber from the plant *Linum usitatissimum,* from which linen cloth is made. Pulp from flax fiber or linen rags makes exceptionally strong, translucent paper.

Gelatin A type of SIZE made from animal tissue or bone.

Glazing The gloss or polish of a sheet of paper and the process by which it is applied.

Hollander A machine, invented in Holland in the late 17th century, for the preparation of rags or fibers for papermaking.

Laid lines The closely spaced lines seen in paper made on a LAID MOLD.

Laid mold A mold whose cover is made of closely spaced, parallel wires or bamboo strips, held in place by more widely spaced, perpendicular chain wires or threads.

Laminating Combining layers of paper by couching one newly formed sheet on top of another to create a single sheet.

Lignin A component of plants that rejects water and tends to decrease fiber-to-fiber bonding in paper. It must therefore be removed before papermaking begins.

Mold A rectangular wooden frame covered with a fine mesh laid or wove wire surface, used for sheetforming.

Parchment A writing surface prepared from the skins of animals, especially sheep and goats.

pH A term used to denote the degree of acidity or alkalinity of a substance.

Post The pile of newly formed sheets alternated with couching felts, ready for pressing.

Pulp The aqueous mixture of ground-up fibrous material from which paper is made.

Size A substance added during beating or after drying to make paper more water resistant. Originally a solution of GELATIN, gum or starch, now various chemical agents.

Spine The back edge of a book where the sections are joined together at the folds.

Template A pattern, usually of thin board or metal plate, from which a similar design can be made.

Watermark (wiremark) A translucent area in a sheet of paper, usually created by attaching a fine wire design to the mold surface.

Wove mold Any mold with a woven mesh surface.

index

acknowledgments

The Publishers would like to thank the many artists and craftspeople who have loaned transparencies of their work.

For advice and the provision of handmade papers, we would like to thank Greig Burgoyne and Atlantis European Limited, 146 Brick Lane, London E1 6RU, art materials suppliers and dealers in fine papers.

For the kind loan of materials, we would like to thank Falkiner Fine Papers, 76 Southampton Row, London WC1B 4AR.

Credits

p.7, Mansell Collection; p.8, Mansell Collection; p.9, Mansell Collection; p.10, Mansell Collection; p.38, embedding showing natural leaf impressions, Coo Geller; p.71, pulp, leaves, and stitching, Amanda Goode; p.88, cast paper clothes, Susan Cutts; p.111, concertina book, Vivien Frank.